Salwa Bouabdallah

The neuroprotective power of Tribulus

Salwa Bouabdallah

The neuroprotective power of Tribulus

The effects of Tribulus extracts in attenuating an Alzheimer's model in zebrafish

ScienciaScripts

Cover image: www.ingimage.com

This book is a translation from the original published under ISBN 978-620-6-70660-1.

Publisher:
Sciencia Scripts
is a trademark of
Dodo Books Indian Ocean Ltd. and OmniScriptum S.R.L publishing group

120 High Road, East Finchley, London, N2 9ED, United Kingdom
Str. Armeneasca 28/1, office 1, Chisinau MD-2012, Republic of Moldova, Europe
Printed at: see last page
ISBN: 978-620-7-66068-1

Summary

Recently, *Tribulus terrestris has attracted* increasing interest due to its pharmacological potential, including its neuroprotective activities. In this study, we explore the neuroprotective effects of a *Tribulus terrestris* extract in a zebrafish *(Danio rerio)* model of scopolamine (SCOP)-induced memory impairment and cerebral oxidative stress. SCOP, an anticholinergic drug, has been used to reproduce fundamental aspects of Alzheimer's disease (AD) in animal models. Fish were treated with ethanolic leaf extract (ELE) of Tt (1, 3 and 6 mg/L) for 15 days. SCOP (100 μM) was administered 30 min before behavioral testing. Molecular interactions of the main compounds identified via UPLC-PDA/MS in Tt fractions with the active site of acetylcholinesterase (AChE) were explored via molecular docking analyses. Terrestrosin C, protodioscin, rutin and saponin C showed the most stable binding. Spatial memory performance was assessed using the Y-maze test, and memory recognition was examined using an object recognition test (NOR). Treatment with Tt extract reversed the impaired locomotion patterns caused by SCOP administration. Biochemical analyses also confirmed Tt's role in inhibiting AChE, enhancing antioxidant enzyme activities and reducing markers of oxidative stress. These results pave the way for future applications of Tt as a natural alternative to treat cognitive disorders.

Key words: Ayurveda; docking; neuroprotection; neurobiology; nutraceuticals; behavioral testing.

Table of contents

Introduction

With increasing life expectancy worldwide, age-related diseases have become a global health priority. The number of people suffering from dementia is expected to reach 152 million by 2050 (Livingston et al., 2020). Alzheimer's disease (AD) is the leading cause of dementia, accounting for 60% to 75% of dementia cases (WHO, 2023). It is characterized by its neurodegenerative nature and manifests as memory-related cognitive impairment and functional decline. The development of AD is mainly governed by genetic factors, but its onset can be linked to exposure to environmental pollutants such as heavy metals and pesticides (Rahman et al., 2020). One of the defining features of AD is the reduction of acetylcholine (ACh) in the brain. This cholinergic hypothesis has been targeted in the treatment of AD.

Approved drugs such as galantamine (GAL), rivastigmine hydrogen tartrate, hierizine-A and donepezil improve cholinergic neurotransmission by inhibiting acetylcholinesterase (AChE) activity (Kareti et al., 2020; Li et al., 2019). These drugs offer symptomatic relief but cannot reverse the progression of Alzheimer's disease and may cause a variety of adverse effects (Dunn et al., 2000; Benninghoff., 2020).

The discovery of new drugs from remedies used in traditional medicine is an attractive approach that is making steady progress. Previous research has focused attention on developing therapies from natural products that could help prevent and manage the neurodegeneration sustained in Alzheimer's disease (Wang et al., 2022). For example, phytochemical compounds such as curcumin and resveratrol have shown neuroprotective activities against key features of Alzheimer's disease, including beta-amyloid accumulation, oxidative stress and neuroinflammation. Some of these compounds have shown high efficacy with low toxicity (Kim et al.,

2010). By therefore, the use of plants in the search for new treatments for Alzheimer's disease is an appropriate strategy (Goozee., 2016; Yan., 2015).

Tribulus terrestris (Tt) is a popular herbal medicine commonly used in traditional Chinese, Indian and Bulgarian medicine (Stefanescu et al., 2022). It is traditionally used for its aphrodisiac, diuretic, anticonvulsant and antihypertensive properties (Chhatre et al., 2014). These traditional uses of Tt have been corroborated by modern research, attracting scientific attention with numerous studies highlighting its pharmacological potentials (Sudheendran et al., 2014; Ara et al., 2023; Phillips et al., 2006). This includes its in vitro anticancer (Bouabdallah et al., 2016), antioxidant, in vitro antileishmanial (Bouabdallah et al., 2018), antidiabetic, immunomodulatory and anti-inflammatory effects. Phytochemical analysis of Tt revealed its abundance in a wide range of structurally diverse compounds, including steroidal saponins, flavonoids, tannins, glycosides, phytosterols and terpenoids (Khalid et al., 2021; Zhao et al., 2023).

Steroidal saponins (Bouabdallah et al., 2023) and flavonoids are the predominant metabolites that confer Tt's pharmacological activities. Over a hundred steroidal saponins have been identified from Tt, including protodioscin, diosgenin, dioscin, gracillin and trillin (Zhu et al., 2017). Recently, these steroidal saponins have been shown to possess neuroprotective properties against Alzheimer's disease (Guan et al., 2022; Yang et al., 2020; Cai et al., 2020).

Zebrafish *(Danio rerio)* have become an essential tool for exploring the effects of various compounds in neurobiology (Kalueff et al., 2014; Khan et al., 2017; Levin et al., 2007)

It has a significant advantage in biomedical research due to its genetic similarity to humans and ease of maintenance (Newman et al., 2014; Johansson et al., 2020).

Zebrafish larvae and adults have been widely used to study neurological processes and disorders (Stewart et al., 2014). The scientific literature confirms the advantage of using zebrafish to study the cholinergic hypothesis of Alzheimer's disease.

Scopolamine (SCOP) is a cholinergic receptor antagonist widely used to develop animal models of cognitive deficits associated with learning and memory (Saleem et al., 2018). It induces features central to Alzheimer's disease, including beta-amyloid accumulation and cholinergic dysfunction.

Literature review

1. Alzheimer

1.1. Epidemiology

Alzheimer's disease (AD) was identified in 1906 by Alois Alzheimer, a German psychiatrist, according to information provided by the Alzheimer Foundation (Alzheimer Foundation, 2019). This doctor characterized this pathology through the study of a patient presenting with various disorders, including orientation problems, memory loss, aphasia, unusual behaviors, as well as psychiatric symptoms such as paranoia and auditory hallucinations. After the patient's death, Dr. Alzheimer obtained her family's permission to examine her brain, where he discovered abnormal deposits between nerve cells.

Because of the early onset of cerebral function impairment in his patient, Dr. Alzheimer referred to this condition as "forgetfulness disease", distinguishing it from senile dementia, which generally occurs after the age of 50. In 1910, the disease was described in a psychiatric textbook as a "particularly severe form of senile dementia", although it has been established that it can sometimes occur before the age of 50.

1.2. Definition

According to current information provided by the Alzheimer's Association (2021), AD is defined as the most common form of dementia, characterized by neurological degeneration. A person is diagnosed with dementia when their loss of memory and other intellectual abilities is severe enough to interfere with daily activities. As the condition develops, it can lead to physical dependence, eventually leading to death if left untreated

(Figure1).

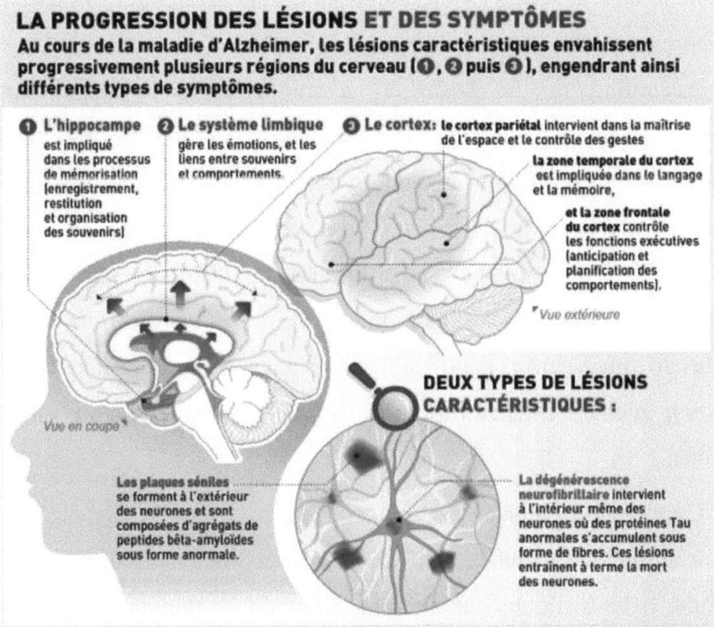

Figure 1: Lesions and symptoms of Alzheimer's disease

Dementia, accompanied by associated psychobehavioral and cognitive disorders (Muller, 2020), is defined as a progressive weakening of cognitive functions, resulting from damage to the central nervous system, and having a significant impact on the individual's social life.

According to the DSM V (Diagnostic and Statistical Manual of Mental Disorders) definition, AD is classified as a neurocognitive disorder, which may be minor or major, characterized by an insidious onset and progressive decline in one or more cognitive functions. This decline may extend over at least 20 years, and must meet the criteria defined to be considered probable AD. The DSM V, the fifth edition of the manual, establishes the different types of mental disorders and their diagnosis, and presents the statistical data available to date. It serves as a reference to ensure greater uniformity among healthcare professionals in the diagnosis of AD.

According to the DSM V, a minor cognitive impairment is characterized by a moderate decline compared with previous cognitive abilities, with no significant impact on autonomy or daily activities. These cognitive declines are not associated with confusional states, and cannot be more satisfactorily explained by other mental disorders such as major depression or schizophrenia, for example. Confusional syndrome is defined as a geriatric emergency, manifesting as an acute disorder of consciousness indicating global, fluctuating and reversible impairment of cognitive functions.

Confusional syndrome is characterized by cognitive impairment that appears abruptly, fluctuates and is often reversible. Individuals with this syndrome have difficulty maintaining attention or thinking clearly. They experience disorientation and fluctuating levels of alertness.

The Psychomédia guide (Psychomédia, 2022) uses the DSM V definition of a major cognitive disorder, manifested by a significant cognitive decline compared with previous performance in one or more cognitive domains, such as complex attention, aphasia (language disorders), agnosia (inability to recognize or identify objects despite intact sensory functions), apraxia (impaired ability to perform motor activities despite intact motor functions) and social cognition. These deficits impact on the person's independence in daily activities, and cannot be explained by other mental pathologies such as major depressive disorder or schizophrenia.

The criteria used to identify a major neurocognitive disorder linked to probable AD mainly include the presence of evidence of AD, established by the existence of a genetic mutation based on family history or genetic test results (Delmas, 2020; Psychomedia, 2022).

In addition, there is a decline in memory and learning, as well as in at least one other cognitive function. Furthermore, cognitive decline is generally progressive, without prolonged periods of stabilization, and there is no significant presence of etiological comorbidities, meaning that there are no other neurodegenerative, cerebrovascular or other neurological, mental or systemic conditions likely to contribute to cognitive decline.

According to data provided by the Vaincre Alzheimer Foundation (VA Foundation, 2021), AD accounts for around 70% of dementia cases in the elderly, with around 1 million people suffering from this disease and related illnesses in France, equivalent to 8% of the French population aged over 65. Projections indicate that by 2050, the number of people with Alzheimer's disease and related disorders in France should reach 1,800,000, representing around 9.6% of people over 65 and 6.2% of the working population.

Worldwide, an estimated 46 million people suffer from neurodegenerative diseases, including AD. Forecasts suggest that by 2040, this number will exceed 80 million cases of neurodegenerative disease worldwide.

Dementia can have a variety of causes, including neurodegenerative diseases such as AD. There are also other conditions linked to AD, such as Lewy body disease, frontotemporal lobar degeneration and neurovascular disorders.

At present, there is no cure for the disease. However, available treatments and ongoing research efforts are aimed at managing the symptoms observed and discovering methods to slow its progression.

1.3. Pathophysiology

Description of the changes observed in the brain

AD is the result of various pathophysiological mechanisms that will be discussed below. According to Poucheret (Pouchret, 2018), the information provided by the Alzheimer's Association and the INSERM website (Inserm. Alzheimer, 2021), the phenomena observed in the brain in connection with this disease have been well described. When examining the brains of people with this pathology, two types of lesions are generally noted: amyloid plaques and accumulations of tau protein in the patient's nervous system (Henstridge et al., 2019; Scheltens et al., 2021).

Within the nervous system, nerve cells establish billions of connections between themselves through networks of neurons. To do so, they need adequate oxygenation to function properly. The presence of amyloid plaques, combined with accumulations of tau protein, leads to neuronal degeneration. Both types of lesion lead to neurofibrillary degeneration, and are generally observed at an advanced stage of the disease (Figure 2). This propagation process, spanning several decades, is responsible for the symptoms observed in the patient. However, changes in the brain can be observed microscopically long before the first symptoms appear (Henstridge et al., 2019; Leng and Edison, 2021).

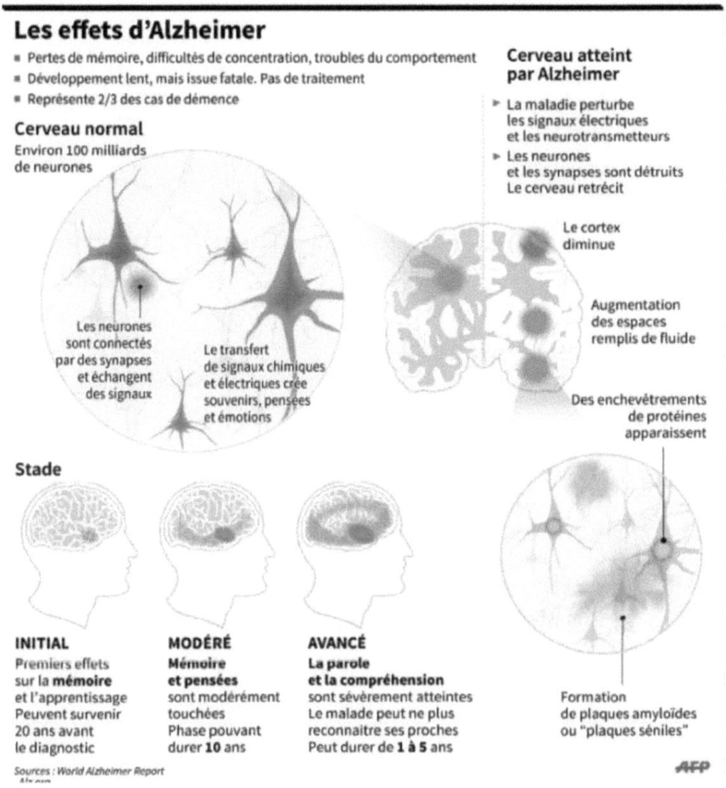

Figure 2: The different stages of Alzheimer's disease

Amyloid plaques, also known as senile plaques, form from deposits of fragments of the protein beta-amyloid or A-beta or Aβ, which lie between the spaces of nerve cells. Initially, this process begins in the hippocampus, before spreading throughout the cerebral cortex, starting with the areas of the cortex involved in memory (Figure 3).

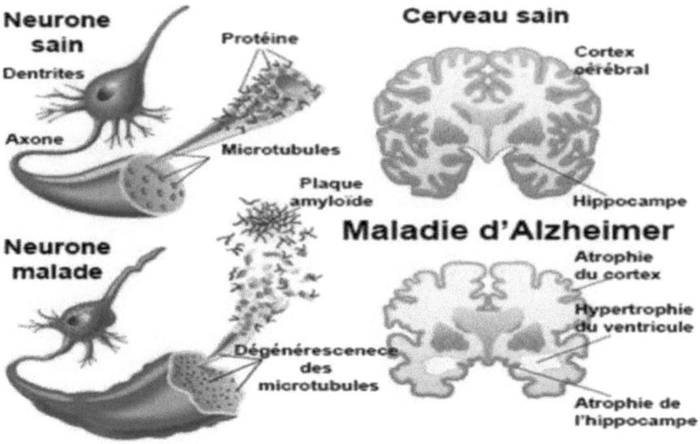

Figure 3. Development of amyloid plaques

The study by Canu and colleagues also highlights the impact of tropomyosin receptor kinase-associated NGF growth factor on the metabolism of the amyloid protein precursor, which contains a region cleavable by the enzyme beta-secretase (Canu et al., 2017). Indeed, NGF binding to the precursor cleavage site, coupled to the receptor, results in increased metabolism of this precursor by the beta secretase enzyme (Fahnestock and Shekari, 2019). This metabolic reaction leads to an increase in beta-amyloid protein production. The discovery of new sites where the pathophysiological mechanisms of AD are expressed could pave the way for the development of new therapies aimed at maintaining normal cholinergic function and normal beta-amyloid protein production (Figure 4).

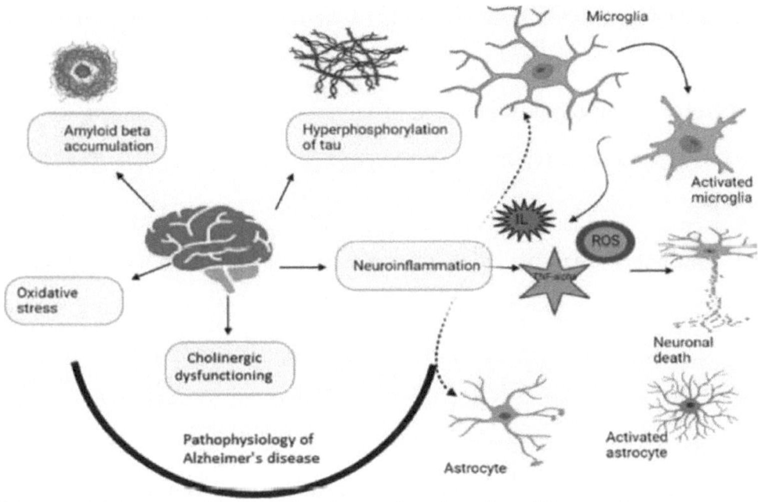

Figure 4. Recent advances in the molecular signalling and treatment of neuroinflammation in Alzheimer's disease.

Another study examined changes in levels of NGF and brain-derived neurotrophic factor (BDNF), two neurotrophins implicated in cholinergic nervous system dysfunction and reduced neuronal plasticity (Iulita et al., 2017). Even before the appearance of amyloid plaques, there is a decrease in BDNF messenger RNA levels, while proNGF messenger RNA expression remains stable, even in advanced stages of the disease. However, NGF precursor protein levels are higher. This disparity between BDNF and NGF levels is seen at all stages of the disease. The increase in proNGF is attributable to a deficiency in a protein responsible for the maturation and degradation of NGF, which stimulates the production of proNGF.

1.4. Physiology of memory

The concept of memory encompasses various types of memory. The patient's emotional context generally exerts an influence on memory formation (Payet, 2018). Firstly, there is very short-term memory, which involves the various sensory cortices. These cortices are located in different regions of the brain, including in particular the occipital cortex, responsible for vision, and the temporal cortex, involved in hearing, olfaction, touch and taste through associative areas. Very short-term memory lasts less than 0.5 seconds.

Short-term memory, also known as perceptual memory, enables us to memorize faces from sight, for example. Also known as working memory, it enables us to associate a sound with an image, or remember a place or a voice (Duffner, 2022). It is involved in the present moment and is used to remember telephone numbers, titles or a series of words, for example. It involves the frontal and prefrontal cortex, the former being responsible for motor skills, while the prefrontal cortex is associated with intellectual, psychomotor and emotional activity. Alterations in these cortical regions can lead to behavioral, mood, intellectual and gestural disorders. The prefrontal cortex must function correctly to enable recall of information in collaboration with the cingulate cortex. This type of memory is an essential prerequisite for memory consolidation. Short-term memory lasts from a few seconds to a few minutes, and its consolidation involves the hippocampus once the information has been acquired in conjunction with the cingulate cortex.

Once the information has been consolidated, different types of long-term memory emerge. These include conscious declarative memory, which is divided into semantic and episodic memory. Declarative memory involves the temporal cortex, hippocampus, amygdala and prefrontal cortex.

Various types of long-term memory have been identified. Episodic memory is linked to our personal history, to the succession of events in our lives. It enables us to recall past moments and anticipate future events. This form of memory is associated with the hippocampus and the lobes of the temporal cortex.

Semantic memory, on the other hand, is based on our culture, school learning and general knowledge (Delmas, 2020). It encompasses knowledge, whether about ourselves or about events (Duffner, 2022). This form of memory is also located in the lobes of the temporal cortex.

In addition, non-declarative, unconscious memory is subdivided into priming, conditioning and procedural activity memories (Payet, 2018).

Procedural memory concerns a person's everyday gestures, habits and practical skills (Delmas, 2020). It involves activation of the cerebellum, striatum and parietal cortex. This form of memory is automatic, and comes into play during activities such as walking or cycling, for example (Duffner, 2022). These gestures mobilize a memory that does not require repeated learning each time they are performed.

In addition, there is emotional memory, which shapes an individual's fears, desires, preferences and attachments (Figure 5).

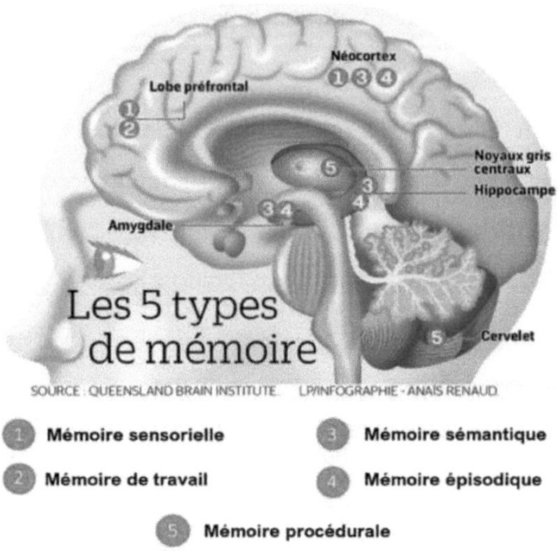

Figure 5. Different types of memory

2. Acetylcholinesterase inhibitors

Rivastigmine EXELON®

Rivastigmine, marketed under the brand name EXELON®, belongs to the anticholinesterase class of psychoanaleptic drugs. Rivastigmine acts as a cholinesterase inhibitor, targeting both acetylcholinesterase and butyrylcholinesterase. Since cholinesterase is an enzyme responsible for the breakdown of acetylcholine, rivastigmine reduces this breakdown, resulting in higher levels of acetylcholine. Since acetylcholine promotes communication between nerve cells, increased levels improve communication between unaffected nerve cells in AD patients.

Rivastigmine is indicated for the symptomatic treatment of AD, both in mild forms, with an MMSE score above 20, and in moderately severe forms, with an MMSE score between 10 and 20 (Poucheret, 2018). This drug is available in capsule or transdermal device form.

ARICEPT® Donepezil

Donepezil, marketed under the specialty name EXELON®, belongs to the anticholinesterase class of psychoanaleptic drugs. Donepezil acts as a cholinesterase inhibitor, giving it a mechanism of action similar to that of rivastigmine. Its aim is to increase acetylcholine levels to facilitate communication between neurons unaffected by Alzheimer's disease.

Donepezil is indicated for the symptomatic treatment of AD, whether mild forms, with an MMSE score above 20, or moderately severe forms, with an MMSE score between 10 and 20 (Poucheret, 2018). It is available in tablet form.

3. Alzheimer's herbal therapy

Herbal therapies can be a fascinating and effective option in the treatment of depression, as a large number of herbal preparations have demonstrated psychotherapeutic activities. The search for new pharmacotherapies from medicinal plants and isolated constituents of plant extracts for psychiatric disorders, including depression, has progressed significantly over the past decade (Zhang, 2004).

For example, a flavonoid-rich fraction obtained from *Monodora tenuifolia* seed extract was able to reduce overall behavioral alterations in rats stressed by forced swimming, and in addition exert protective effects against induced oxidative stress, supporting its antidepressant effect (Ekeanyanwu and Njoku, 2015).

In another study, the methanolic extract of the species *Byrsonima crassifolia (L.)* Kunth (Malpighiaceae) revealed antidepressant activity in the forced swim test, and the antioxidant flavonoids rutin, quercetin and hesperidin may be involved in the antidepressant effects of *B. crassifolia (L.)* Kunth (Ahmadi and Shadboorestan, 2016; Herrera-Ruiz et al., 2011; Kanimozhi et al., 2017; Sun et al., 2017).

Flavonoids are a broad class of secondary metabolites abundant in plants and various foods. They have been identified in a variety of fruits and vegetables and confer color, flavor and aroma, as well as nutritional and health benefits. Polyphenolic flavonoids are the most effective functional ingredients with biological activities. Many flavonoids possess antioxidant and antidepressant activities (Guan and Liu, 2016; Kanimozhi et al., 2017; Sun et al., 2017 ;).

It is widely reported that oxidative stress plays a critical role in the development of various diseases (Liu et al., 2017), including psychopharmacological disorders (Salim, 2017). Indeed, the link between oxidative stress and depression has been studied and discussed in some reviews (Filipović et al., 2017; Maurya et al., 2016; Vaváková et al., 2015) (Figure. 6).

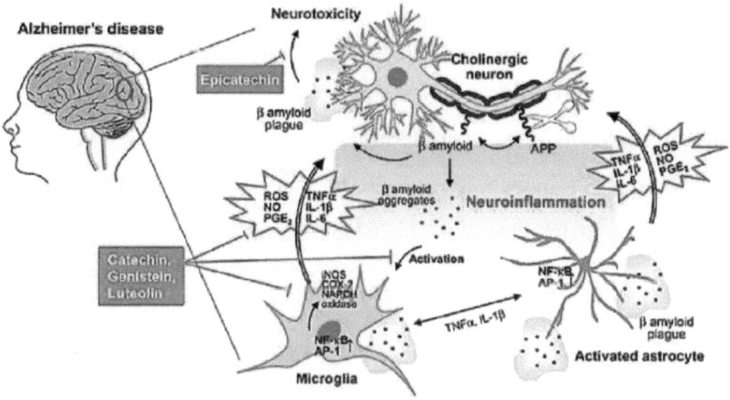

Figure 6. Effect of flavonoids on neuroinflammation in Alzheimer's disease.

4. The Zebra Fish

4.1 Application of zebrafish in neuroscience research

Native to Southeast Asia, the zebrafish *(Danio rerio)* has established itself as a widely used model organism in biomedical research (Figure 7). Several advantages of using this species in biomedicine include significant physiological and genetic similarities to mammals, external fertilization, rapid developmental processes, transparency of embryos and larvae, ease of genetic and experimental manipulation, as well as cost-effectiveness and space efficiency (Cachat et al., 2010; Collier and Echevarria, 2013; Grossman et al., 2010; Parker et al., 2013; Stewart et al; Stewart et al., 2012).

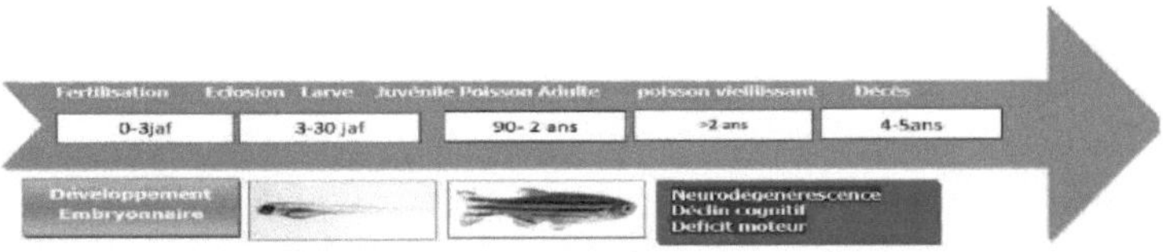

Figure7. Zebrafish *(Danio rerio}* adult and larval stages: Zebrafish life cycle, from pre-hatching embryonic period (0-72 haf) to post-hatching stages, including larval (3-29 dpf), juvenile (30-89 dpf), adult (90 dpf-2 years) and old (> 2 years) stages.

The statement suggests that there have been extensive reviews of the strengths and limitations of zebrafish models in biomedical research, particularly their relevance to neuroscience (Ablain and Zon, 2013; Ahmad et al., 2012; Brennan, 2011; Gerlai, 2012; Kalueff et al., 2014; Steenbergen et al., 2011). In conjunction with mammals and well-known invertebrate model organisms, both larval and adult zebrafish are widely used in central nervous system (CNS) research (Drew et al., 2012; Miller et al., 2013; Pippal et al., 2011) and in the study of various brain disorders. However, wider adoption of zebrafish models in neuroscience research faces challenges due to limited familiarity with these models and their phenotypes among non-fish neuroscience laboratories.

For a fuller understanding, readers are referred to the details provided in reference (Kalueff et al., 2014). Recognizing the potential of zebrafish in translational neuroscience (Brennan, 2010; Kalueff et al., 2014; Miller et al., 2013), this work has explored its growing importance in investigating the causes of brain disorders and experimentally manipulating these conditions (Buske et al., 2011; Buske and Gerlai, 2011; Gerlai, 2011; Kalueff et al., 2013).

Several advantages of using this species in medical biology include high physiological and genetic homology with mammals (70-75%), external fertilization, significant advances have been made recently in the development of zebrafish models relevant to a range of conditions, including autism, sleep disorders, cognitive deficits, depression, psychoses and addiction (Brennan, 2011; Cachat et al., 2013; Griffiths et al., 2012; Grossman et al., 2010; Kyzar et al., 2013; López et al., 2008;Parker et al., 2012; Neelkantan et al., 2013; Singh A, et al., 2013; Stewart et al., 2014; Ziv et al., 2013).

In summary, accumulating evidence suggests that the zebrafish is rapidly emerging as a lead organism in translational neuroscience and bio-psychiatry research. It effectively complements both rodent and clinical models, contributing to the understanding of almost all major brain disorders (Brennan, 2010; Kalueff et al., 2014; Kalueff et al., 2014; Stewart et al., 2013).

4.2 An effective animal model for studying Alzheimer's disease

The zebrafish is gaining in importance as a valuable model organism for the study of a range of CNS disorders, including AD. Given that AD represents a major contributor to dementia in humans, there is an urgent need to understand the mechanisms underlying this neurodegenerative condition. In this context, it is absolutely necessary to create new animal models to explore the fundamental neurodegenerative processes of AD.

The review by Stewart et al, (Stewart et al., 2014) explores the current state of the use of zebrafish as a model of AD, examining the rationale behind the employment of this experimental approach in CNS research. The review encompasses various strategies used to induce AD-like pathology in zebrafish, providing insights into their potential as a model organism for understanding the CNS-related aspects of AD. It explores various strategies involving interventions to induce damage in the zebrafish brain, targeting key neurotransmitter systems such as cholinergic, glutamatergic or GABAergic circuits.

This review encompasses models of familial AD, which are based on the amyloid precursor protein (APP) and presenile processing pathways, as well as those focusing on Tau protein Thyperphosphorylation. Genes linked to sporadic AD, such as apolipoprotein E, are also taken into consideration. Finally, it can be concluded that the use of zebrafish in AD is currently in full development, underlining the need to refine transgenic models to establish this organism as an optimal model for AD research. In addition, there is a need to better understand the zebrafish brain and to characterize more precisely the damage resulting from alterations in key neurotransmitter systems.

5. Tribulus (Tribulus terrestris L.)

5.1 History

Tribulus terrestris (Tt) is a plant that has been used for hundreds of years in Chinese and Indian (Ayurvedic) medicine. Several medical systems have been used in Asian countries since prehistoric times. The Ayurvedic system, practiced in Sri Lanka, India and other parts of Asia, is the most popular.

In India, 700 BC, a great Ayurvedic physician named Charaka wrote a medical treatise based on earlier medical texts. In this treatise, Charaka clearly describes a plant *(Tribulus terrestris L.) with* diuretic, aphrodisiac, tonic, rejuvenating and strengthening properties.

Later, Chinese medicine added to its knowledge of Tt and named it "the herb that extinguishes wind and stops trembling". In China, "the wind" corresponds to heart disease (strokes, hypertension).

In 1847, a scientist named Fritzsche extracted a chemical molecule called harmine from *Peganum harmala.* Harmine is one of the active compounds in Tt and is used as a stimulant and aphrodisiac.

On the other hand, explorers in South Africa found that the ingestion of plants containing harmine had an effect on the mind and consciousness; this psychotropic substance was subsequently used in the treatment of Parkinson's disease.

Since then, numerous studies have been carried out to determine Tt's chemical composition and discover its pharmacological benefits.

5.2 Classification

Tribulus is classified as follows (Linnaeus)

- Spermatophytes
- Angiosperms
- Dicotyledons
- Disciflores series
- Order of Geraniales
- Family Zygophyllaceae
- Genus *Tribulus*
- *Terrestris* species.

Botanical study

Tribulus terrestris is a rather unusual species, with stems that lie flat on the ground, often spreading out in a radiating pattern from the point where the main root is embedded. It is found in wastelands, fields and along roadsides, but can also grow in sandy soil. Its small yellow flowers, isolated from one another, open from April to October.

This species is remarkable for its leaves, which have 8 to 16 oval, opposite leaflets arranged in two rows, to the right and left of the common petiole, with no terminal leaflet at the apex of the leaf; below the leaflets, the leaf petiole is quite short (Figure 8). The flowers, borne on slightly elongated peduncles, are sometimes placed in the leaf axils, sometimes on the bifurcations of the stem, sometimes opposite the leaves; in the latter case, the stem is formed by sections of successive branches. The sepals are oval; the petals are about one-and-a-half times the length of the sepals; the 5 stigmas are radiating and reflecting on the thick, very short style (Bonnier, 1911).

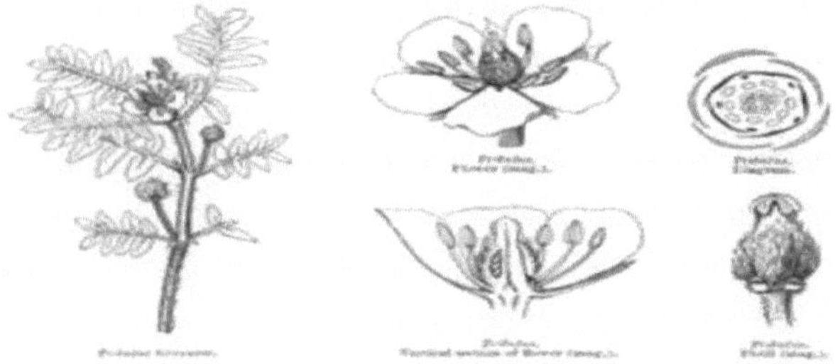

Figure 8. Microscopic section of *Tribulus terrestris* reproductive organs

The fruit, which is more or less hairy, has a very special shape; when ripe, it separates into 5 parts, each with a more or less elongated spine towards the top, and two shorter spines further down (which sometimes don't develop) (Chadefaud, 1960).

Each of these 5 parts of the fruit detaches without opening and falls with the 2 or 3 seeds it contains (Figure 9), these seeds are placed one on top of the other, embedded in small logettes which are separated from each other by transverse partitions.

Figure 9. Morphological appearance of *Tribulus terrestris* seeds

It's a more or less hairy plant, sometimes whitish, sometimes more or less green, depending on how many hairs it's covered with. The species is an annual with a developed main root.

5.3 Geographical distribution of the plant

Tribulus terrestris is native to Europe, and generally grows in arid climates and sandy soils along roads and highways. It is mainly found in the tropics and in all hot regions of the world.

-Traditional use

Outside North Africa, it grows in Europe, Western Asia, India, the Canary Islands, Senegal, the Cape of Good Hope and the Comoros Islands. It is grown in western France as far as the Loire Atlantique, as well as in America (Bonnier, 1911).

This plant is used for a variety of purposes in many countries: India, North Africa, Bulgaria, China, South Korea, Europe, Kuwait, Nepal, Pakistan, Peru, Thailand, Tanzania and Turkey (Bonnier, 1911).

-Use by country

In **India,** the whole plant is used in cases of leucorrhoea (or white discharge: mucous or muco-purulent discharge from the vulva), while its aqueous extract is an aphrodisiac and treats kidney stones. The aqueous extract of the root is an emmenagogue, i.e. it induces or regulates menstrual flow. The fresh seed, combined with honey, is tonic, improves skin vitality and shine, reduces wrinkles and treats jaundice. Dried fruit powder is used as a narcotic; too much fruit can cause delirium. Fresh fruit juice is used for urinary disorders. Root powder treats gonorrhea (urethral discharge). Aqueous fruit extract is a tonic, diuretic and aphrodisiac. Infused, it acts as a uterine tonic. The fruit is used in cases of impotence, urolithiasis, gonorrhea and kidney disease. *In cases of* acute newborn debility, it is used in association with *Curculigo orchiodes* and *Echinops echinatus* roots.

In **North Africa,** it is used as an antidiarrheal, stimulant and aphrodisiac.

In **Bulgaria,** the dried aerial parts are taken orally to increase spermatogenesis.

In **China,** the aqueous extract of the aerial parts is administered in doses of between 7 and 10 mg, and is indicated in cases of spermatorrhea (involuntary emission of semen). Aqueous extract of dried seeds is used for liver disease, *while* the fruit is used for eye disorders, edema, abdominal distension and leucorrhoea (white discharge).

Dried fruit extract is indicated externally for skin hyperpigmentation and as a beauty product (cosmetics). The fruit is abortifacient. Dry powder combined with butter and honey improves longevity (Ivan, 2001).

We note that, depending on the nature of the drug, the uses can be very different. Each part of the plant has a different composition, both in terms of product quality and quantity. It is this variation in composition that confers a more specific activity of one part of the plant on a given disease. For example, the fruit is more specific for kidney disease, the root for liver disease, the seed for skin disease and the leaf for diarrhoeal disorders (Samuelsson, 1993).

On the other hand, the same plant part can be used in different forms (dry, fresh, extract, powder) and for very different indications. Processed products have a different composition from that of the fresh drug, depending on the transformations carried out (heat, extraction). For example, dried fruit has narcotic activity, fresh fruit treats urinary disorders and aqueous extract is diuretic and aphrodisiac (Samuelsson, 1993).

There are also variations in action depending on the solvent used (water, ethanol, methanol, chloroform), as these different solvents will result in different compounds. As a result, the extracts obtained will differ in composition and activity.

We note that, depending on the country, certain uses of Tt are identical. Some countries are very far away, and were unable to communicate their experiences at a time when the plant was commonly used in traditional medicine. So the more countries use or used the drug for a particular indication, the more likely it is to be effective. From the above table, we deduce that diuretic and aphrodisiac actions are the most frequently encountered and therefore the most plausible, followed by tonic, stimulant and anti-impotence actions. Aqueous extracts of fruit, leaves and roots are diuretic. Aqueous extracts of roots and fruits are aphrodisiacs. Aqueous fruit extract is tonic and treats impotence (Ivan, 2001).

5.4 Chemical composition of aerial parts

The main active compounds isolated from aerial parts are :

Steroidal saponins (Ivan et al., 2001; Dinchev et al., 2008)

- glucopyranosyl galactopyranes
- ruscogenin heterosides, hecogenin, diosgenin, gitogenin, neotigogenin, chlorogenin, tigogenin, terrestroside F
- protodioscin glycosides
- protodioscin, dioscin

Flavonoids (Bhutani et al., 1969; Dinchev et al., 2008).

- Kaempferol
- Kaempferol-3-O-glucoside

- Kaempferol-3-O-rutinoside
- *Kaempferol-3-O-*(6ʿ’ -O-E-p-coumaroyl)-β-Dglucopyranoside
- Tribuloside
- *Isorhamnetine-3-O-*(6ʿ’ -O-E-p-coumaroyl)-β-Dglucopyranoside
- Quercetin-3 -O-β-D-glucopyranoside
- Rutin
- Amino acids and proteins
- Phytosterol

-Macronutrients: potassium, calcium, sodium, phosphorus

-Indole alkaloids: Harman, harmine.

5.5 Main pharmacological properties

A number of biological and pharmacological studies have been carried out on the beneficial effect of Tt, the main properties of which are listed below.

Regulation of NO and androgen receptors

- Tt stimulates libido in rats (Bonnet, 2003)

- Tt increases the number of NADPH-diaphorase neurons (NO synthase-dependent enzymatic activity) and androgen receptors in the hypothalamus (Gauthaman and Adaikan, 2005).

- Protodioscin isolated from Tt enhances the smooth muscle relaxant effect of male genitalia, probably by increasing nitric oxide (NO) release from the endothelium and nerve endings (Adaikan, 2005). This compound also increases intracavernous pressure (Gauthaman et al., 2002; Gauthaman et al, 2003).

- A positive effect of Tt extract on spermatogenesis (sperm count and motility) has been demonstrated in humans (Protich, 1983).

The saponin fraction of Tt promotes ostrogenic functions and fertility (Bonnet, 2003).

Diuretic and urinary anti-lithiasis agent

Al-Ali's study showed that Tt administration in vivo increases urinary excretion and natriuresis (Al-Ali et al., 2003).

Tt is also said to have a preventive effect on kidney stone formation, as its administration to rats with hyperoxaluria reduces urinary oxalate excretion and increases glyoxylate excretion. This is due in particular to a reduction in the activities of the hepatic enzymes that synthesize oxalate, and a normalization of that of renal lactate dehydrogenase (Sangeeta, 1994).

Hypotensive effect

In vivo, Tt reduces blood pressure by decreasing ACE (Angiotensin Converting Enzyme) activity (Sharifi et al., 2003).

Cytotoxic effect

8Aqueous fruit extract (250µg/ml and 500µg/ml) of Tt exhibited cytotoxic potency against liver cancer cell lines (HepG2) by inducing cell apoptosis (Kim et al., 2010).

Bouabdallah et al, studied the anti-tumor potential of Tt grown in Tunisia: Different extracts (ethanolic, butanolic, aqueous...) of Tt (leaves seeds and roots) were evaluated against cancer cell lines (OVCAR, IGROV, Hep2 and K562). Thus, the n-butanolic extract of leaves is endowed with the best antitumor effect of the order of 94.76 ± 1.52% at 50 µg/mL (Bouabdallah et al., 2016).

Anti-leishmanial effect

Bouabdallaah et al. investigated the antileishmanial potential of different extracts isolated from Tunisian *Tribulus.* Their results showed that the n-butanolic extract (saponin-rich fraction, isolated from leaves) exhibited the best antileishmanial effect against the pathogenic parasites Leishmania *L. major* (GlC94) and *L. infantum* (LV50) assessed in vitro by an MTT assay (Bouabdallah et al., 2018).

Materials and methods

1-Extraction of plant matter

Tribulus terrestris (Tt) was collected from the Tunisian flora, from plants growing spontaneously in the Elhawaria region (Ain Halloufa). Samples were collected during the fruit ripening phase in September 2020. Harvested plants were identified according to Pottier-Alapetite criteria, and confirmed according to methods previously described by our botanical colleagues as mentioned in our publication (Bouabdallah et al., 2016). The extraction protocol was carried out as previously reported by Bouabdallah et al, with slight modifications (Bouabdallah et al., 2016, 2018). Fresh Tt leaves were shade-dried for two weeks and then reduced to a very fine powder (Figure10). One hundred grams of the powder obtained was extracted using 70% v/v ethanol (reflux at 80°C, 3 x 1000 mL x 2h). The resulting solutions were combined and concentrated to a small volume under vacuum at 70°C.

Figure 10. Different stages in the preparation of *Tribulus terrestris* grown in Tunisia (Bouabdallah, 2020)

2. Chemical analysis of investigated extract

- Ultra-high performance liquid chromatography - Photodiode array

detection

A Thermo Fisher UltiMate 3000 system coupled to a photodiode array (PDA) detector was used for mass spectrometry (MS) analysis. The mobile phase comprised a gradient A (acetonitrile with 0.1% phosphoric acid); and B (0.1% phosphoric acid) as follows: 0-4 min. 10-15%, next 4 min. isocratic at 15%, 8-15 min. 30%, 15-18 min. 40%, 18-22 min. 55%, then for the next 3 min. the system returned to initial conditions. The column used was Luna Omega 5 μ Polar C18 (100A, 150 × 4.6 mm). Sample injection was 2 μL at a flow rate of 0.8 mL. Detection was performed at seven different wavelengths between 220 nm and 800 nm, taking into account the four detection bands at 245 nm, 280 nm, 330 nm and 521 nm. These wavelengths represent the maximum absorption bands for flavonoids, phenolic acids and anthocyanidins, respectively.

Each peak was identified on the basis of available UV spectra and library data, and the identifications were compared to standards. Calibration curves were established for several standards, including catechin, epicatechin, ellagic acid, caffeic acid, p-coumaric acid, cinnamic acid, chlorogenic acid, ferulic acid, luteolin, apigenin, kaempferol, epigallocatechin, quercetin, rutin, quercetin-3-O-arabinoside, apigenin-7-O-glucoside, luteolin-7-O-glucoside and ecdysone, for quantification of the compounds identified.

A system comprising a Transcend TLX-1 Vanquish Flex chromatograph coupled to an Orbitrap Exploris 480 mass spectrometer was used to further evaluate the chemical composition of the extract studied by liquid chromatography coupled to mass spectrometry (UPLC-ESI/MS). Full-scan evaluation was carried out on a Hypersil Gold C18 column (50 x 2.1 mm, 1.9 μm) thermostated at 40°C. Separation of active compounds was carried out with a mixture of

gradient A (acetonitrile with 0.1% phosphoric acid) and B (0.1% phosphoric acid) over 30 minutes, from 7% to 93% A in B.

Flow rate was 0.3 mL/min. Positive and negative electrospray ionization (ESI) was used for mass spectrometric detection. Acquired spectra were recorded in a mass range from *m/z* 100 to 1500 at a frequency of 5 Hz, using a positive voltage of 4.5 kV, a negative voltage of 3.0 kV and an ionization temperature of 350°C. Aliquots of 5 μL of each sample were injected automatically. Integration and detection were performed using Compound Discoverer 3.2 software for evaluation of targeted and non-targeted modules. Peaks with a score of at least 7.4 were considered and checked against the Thermo *m/z* Vault and Chem spider databases.

3. Molecular docking

With the help of the MOE-Dock 2019.09 program from TEnvironnement Opérationnel Moléculaire (MOE), a docking analysis was carried out. Using the manufacturer's key, sketches of the structures of the selected molecules were generated. The MMFF94x force field, a standard feature of the MOE program, was then used to reduce the energy level of these compounds. Then, using a conformer search, 3D conformers of these compounds were obtained. AChE was obtained from the Protein Data Bank (PDB ID: 4EY7) (Cheung et al., 2012), and was imported into MOE.

Water molecules were removed, and missing hydrogens were added to generate the necessary ionization conditions of the protein structure. The location of the active site was determined using the MOE Alpha Site Finder with its default parameters. The resulting alpha spheres were used to construct the artificial atoms of the inactive site. The docking process was carried out using MOE's "Docking" component, following the standard docking procedure. The five best poses, ranked according to the London Dispersion Grid (dG) criterion, were saved, further reducing the use of the Merck 94x molecular force field (MMFF94x) within the enzyme. The dG Generalized-Born Volume Integral/Weighted Surface Area (GBVI/WSA) ranking utility was then used to assign scores to the generated poses. Compound poses with the highest scores were selected. We used the Biovia Discovery Studio 2020 viewer to examine the representation of protein-ligand interactions in the active site of the complex (Abo-Elghiet et al., 2023).

4. Ethical approval

The maintenance and treatment of the fish were carried out in accordance with the Recommendation of the European Commission (2007), Directive 2010/63/EU of the European Parliament and of the Council of 22 September 2010 on the accommodation, care and protection of animals used for experimental and other scientific purposes. This protocol was approved by the Ethics Committee of the Faculty of Biology in Iasi. All procedures were carried out in compliance with legal guidelines and "Animal research: report of in vivo experiments (ARRIVE)", while ensuring minimal use of zebrafish.

5. Fish farming

A total of 100 wild-type short-finned strain zebrafish were obtained from Pet Product S.R.L. (Bucharest, Romania). This included both female and male zebrafish in a 1:1 ratio, aged 3 to 4 months, measuring 3 to 4 cm in length, and weighing between 0.3 and 0.5 g. The zebrafish were acclimatized for a period of two weeks in the animal facility of the Faculty of Biology in Iasi (Romania). They were housed in three 70 L aquariums and maintained in a recirculation system supplying well-ventilated, dechlorinated water at a controlled temperature (26°C ± 2). The photoperiod was set to 14:10 h (light:dark cycle). Water quality parameters were maintained at constant levels (pH = 7.5, dissolved oxygen at 7.20 mg/L, ammonium concentration <0.004 ppm, and conductivity level of 500 μS).

6. Pharmacological treatment and group allocation

Zebrafish were divided into the following groups: Three pretreatment groups with Tt comprising three concentrations (1, 3 and 6 mg/L); the control group (1% DMSO was added by immersion); the SCOP group (SCOP, 100 μM, Sigma-Aldrich, Darmstadt, Germany); and the galantamine (GAL) group (GAL, 1 mg/L, Sigma-Aldrich, Darmstadt, Germany) serving as a positive control in the Y-maze and NOR tests. Doses of SCOP, GAL and plant extract were selected on the basis of previous studies (Dumitru et al., 2019). Tt (1, 3 and 6 mg/L) was administered by immersion once a day in a 6 L glass for 1 h, while SCOP (100 μM) was administered 30 min before behavioral testing.

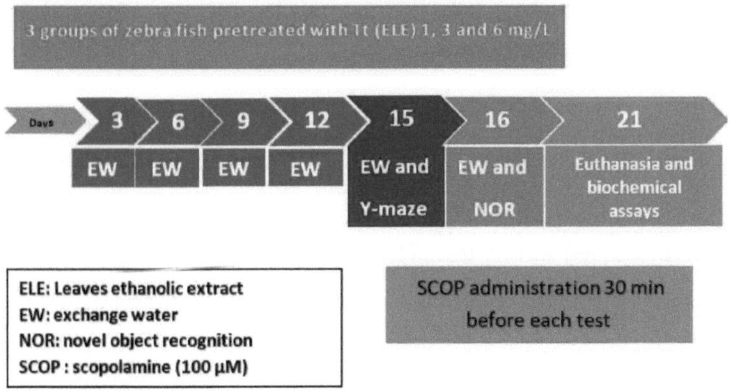

Figure 11. Experimental design procedure for treatment administration, behavioral studies (NOR and Y-maze tests) and biochemical assays (Bouabdallah et al., 2024)

6. Behavior assessment

A Logitech HD Webcam C922 Pro Stream (Logitech, Lausanne, Switzerland) recorded zebrafish behavior. The recorded videos were reviewed using ANY-maze® version 6.3 software (Stoelting CO, Wood Dale, IL, USA).

6.1. Y maze test

The novelty response in zebrafish was assessed through the Y-maze task (Cognato et al., 2012). Spatial position in the Y-maze test served as a memory indicator. Zebrafish were individually trained in a Y-shaped glass aquarium (3 L) with three arms (25 × 8 × 15 cm), which were randomly assigned as the "start" arm (always open), the "novelty" arm (open during the trial) and the "permanently open" arm.

During the first training session (5 minutes), the fish were individually positioned in the starting arm, and the novelty arm was closed. After one hour, the second 5-minute training session began, and the fish were again positioned in the starting arm, but this time the novelty arm was opened.

Total distance traveled (in meters), time spent in the novelty arm (in % of total time in the arms), and rotation angle (in degrees) were the behavioral parameters studied in this test. The recorded videos were evaluated using ANY-maze® software (Stoelting Co., Wood Dale, IL, USA).

6.2. New object recognition test

The novel object recognition (NOR) test is generally used to assess behavioral memory capacity in zebrafish. Zebrafish were briefly subjected to a 5-minute acclimatization period in the new tank (30 × 30 × 30 cm filled with water over 6 cm) for three consecutive days with no object present (Stefanello et al., 2019; Gaspray et al., 2018). During the training phase (day four), zebrafish were introduced to two similar objects for 10 minutes. One hour after the start of the training phase, one of the two similar objects (FO: familiar object) was arbitrarily replaced by a new object (NO) for 10 minutes (test phase). Percentages were determined according to the following formula: [NO exploration time / FO exploration time + NO exploration time × 100]. Behavior was monitored and analyzed using ANY-maze® software (Stoelting Co., Wood Dale, IL, USA).

7. Biochemical analysis

Collected brain tissue was gently homogenized (ratio 1:10, w/v) in cold 0.1 M potassium phosphate buffer (pH 7.4) containing 1.15% KCl. Homogenization was carried out for 1 minute at 1000 rpm using the Mikro-Dismembrator U mill (Sartorius, New York, NY, USA), equipped with 3 mm diameter magnetic beads (Sartorius Stedim Biotech GmbH, Goettingen,

Germany). The resulting homogenate was then centrifuged for 15 minutes at 14,000 rpm. The supernatant was collected for further analysis. This included measurement of total soluble protein content, and assessment of the activities of specific enzymes such as AChE, superoxide dismutase (SOD), catalase (CAT), glutathione peroxidase (GPX), glutathione (GSH), malondialdehyde (MDA) and carbonylated proteins. All procedures were carried out in compliance with applicable rules and regulations.

AChE activity was estimated in brain supernatant using the Ellman method (Ellman et al., 1961). SOD activity was measured according to the Winterbourn protocol (Winterbourn et al., 1975). Catalase activity was analyzed using the Sinha method (Sinha et al., 1972). GPX activity was estimated using the method of Sharma and Gupta (Sharma and Gupta, 2002). Carbonylated protein levels were assessed according to the protocol described by Oliver et al. (Olivier et al., 1987) and modified by Luo and Wehr (Luo and Wehr, 2009). Lipid peroxidation (malondialdehyde level, MDA) was assessed using the Ohkawa method (Ohkawa et al., 1979). Protein content in brain supernatant was quantified using the Bradford method (Bradford et al., 1976).

8. Statistical analysis

Data were presented as means ± standard error of the mean (SEM) and were statistically analyzed using a one-factor analysis of variance (ANOVA), followed by Tukey's post hoc multiple comparison tests. Data were interpreted using GraphPad Prism v8.3 software (La Jolla, CA, USA). A statistically significant difference was considered for $p < 0.05$.

Results

1. Phytochemical analysis of *Tribulus terrestris*

Chromatographic results indicated the presence of several compounds in the sample, as shown in Figure 12. The Tt extract was analyzed by UPLC-PDA to determine its chemical profile. The results revealed the presence of various flavonoids and saponins, including epigallocatechin, apigetrin, rutin, quercetin, luteolin, cynaroside, caffeic acid, trillin, hecogenin, terreside B, trillarin, protodioscin and saponin C (Figure 12).

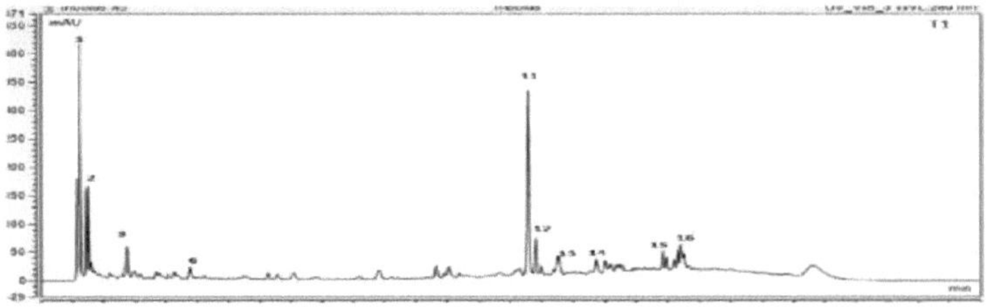

Figure72. UPLC-PDA chromatogram for a *Tribulus terrestris* sample (ELE: ethanolic leaf extract) at 280 nm.

Compound identification was confirmed by UPLC-EIS/MS, based on literature review and fragmentation patterns compared to similar compounds in the online database using Compound Discover 3.2 software for evaluation including targeted and non-targeted modules. Peaks with a score of at least 7.4 were considered and checked against the Thermo m/z Vault and Chem spider databases. In this respect, Tt metabolite profiling led to the provisional identification (Table 1) of a wide range of metabolites represented by phenolic acids, flavonoids and saponins. Molecular mass ion peaks at m/z 180.04225 and 196.05823, corresponding to the suggested molecular formulae C9H8O4 and C9H8O3 $[M+H]^-$, correspond to caffeic acid and hydroxycinnamic acid. Molecular ion peaks at m/z 285.0409, 301.0353,

356.07451 and 609.1482 $[M+H]^-$, respectively, for the formulae

C15H10O7, C15H10O6, C15H14O7 and C27H30O16 were identified as kaempferol, quercetin, epigallocatechin and rutin $[M+H]^-$. While molecular ion peaks at *m/z* 913.48023, 915.4590, 738.05805 and 1049.2 $[M+H]^-$, respectively, for molecular formulae
C51H84O12, C45H12O15, C39H62O13 and C51H84O22 correspond to a C saponin, a C terrostrosin, a trillarin and a protodioscin (Table 1).

Table 1. UPLC-ESI/MS peak designations for metabolites in *Tribulus terrestris* (ELE).

No	Formula	Transformations	Error (ppm)	$[M-H]^{-1}$ Experimental	$[M-H]^{-1}$ Theoretical	Identification
1	C9H10O7	Hydration, Oxidation	-0.03814	180.04225	179.03498	t-caffeic acid
2	C9H8O3	Hydration, Oxidation	-0.02587	198.0402	199.0606	Hydroxycinnamic acid
3	C21H20O11	Reduction	-0.38036	196.05823	195.05095	Cynarosia
4	C15H10O7	Reduction	0.17143	301.0353	302.24	Quercetin
5	C15H10O6	Nitro Reduction	-0.21529	285.0409	286.24	Kaempferol
6	C15H10O6	Nitro Reduction	0.28338	285.04120	286.24	Luteoline
7	C15H14O7	Hydration, Oxidation	0.45039	356.07451	306.27	Epigallocatechin
8	C21H20O10	Reduction	0.18654	435.5	432.4	Apigetrin
9	C27H30O16	Hydration	-1.45637	609.1482	610.1084	Rutin
10	C45H72O17	Reduction	-4.4	915.4550	915.4590	Terreside A
	C27H42O4	Reduction	-0.38036	255.08642	430.30764	Hec ogenine
11	C39H62O14	Desaturation, Nitro Reduction	0.3610	755.262	754.901	Terreside B
12	C45H12O15	Desaturation, Nitro Reduction	-0.38036	918.4232	915.4590	Terrestrosin C
13	C39H62O13	Desaturation, Nitro Reduction	0.95112	738.05805	738.05077	Trillarin
14	C51H84O22	Oxidation, Nitro Reduction	-0.10032	1047.5413	1049.2	Protodioscin

15	C33H52O8	Desaturation, Nitro Reduction	2.37949	560.22	576.761	Disogluside - Trillin

2. Molecular Docking Study

AChE is a crucial enzyme that inhibits neurotransmission by hydrolyzing ACh at the synapse. AChE is mainly found at neuromuscular junctions and cholinergic synapses in the central nervous system (CNS), where it promotes the hydrolysis of ACh to choline and acetate, thus terminating ACh-mediated synaptic transmission with high catalytic efficiency. The structure of AChE consists of three central binding regions: the peripheral aromatic site (PAS), the catalytic active site (CAS) and the throat (Moreira et al., 2022). The catalytic triad of Ser203, His447 and Glu334 is present on the bottom of the CAS-containing groove. In addition, the central groove connects CAS and PAS and extends beyond the enzyme (Lio et al., 2022). The majority of the inner surface is composed of aromatic amino acids (Phe295, Phe338 and Tyr337). The PAS at the throat entrance is mainly composed of five residues: Asp74, Tyr72, Tyr124, Tyr341 and Trp286 (Lio et al., 2022). The docking scores of the 15 compounds against AChE ranged from -11.22 to -24.68 kcal/mol, compared with -14.62 kcal/mol for donepezil (co-crystalline ligand) (Table2). With docking energies below -24 kcal/mol, terrestrosin C, protodioscin, rutin and saponin C were the most stable docked compounds. The order of highest docking scores was as follows: Rutin > Saponin C > Protodioscin > Terrestrosin C > Trillarin > Epigallocatechin > Terreside B > Cynaroside > Disogluside, Apigetrin > Kaempferol > Quercetin > Hecogenin > Luteolin > Donepezil > Caffeic acid.

Table 2. Binding energy of compounds identified in *Tribulus terrestris* with the AChE enzyme.

	Compounds	Docking energy scores in kcal/mol
	Donepezil (Co-crystal ligand)	-14.62
1	t-Caffeic acid	-11.22
2	Disogluside (Trillin)	-18.60
3	Apigetrin	-18.60
4	Cynaroside	-18.70
5	Terreside B	-19.98
6	Terrestrosin C	-24.11
7	Trillarin	-23.42
8	Protodioscin	-24.58
9	Epigallocatechin	-20.62
10	Rutin	-24.68
11	Hecogenin	-14.71
12	Saponin C	-24.63
13	Quercetin	-15.39
14	Kaempferol	-15.48
15	Luteoline	-14.65

3. Effect of Tt extract on locomotor activity

Digital videos were recorded to observe the movement of zebrafish in the water, vertically or horizontally, over a period of time. A trajectory drawn in specialized software was used to automatically track the zebrafish's swimming style. Representative swimming behavioral parameters were analyzed, including average total distance covered, speed or turn angle (Brinza et al., 2020, 2022; Capituna et al., 2020). A basic approach to analyzing zebrafish movement is to count the total number of lines crossed by the fish. The memory improvement observed in the Y maze test can be attributed to locomotor activity, which is assessed by the number of entries into the maze arms. Groups pre-treated with SCOP showed significant variations, indicated by $p < 0.05$. As shown in Figure 13, the control group exhibited regular swimming behavior in the Y maze tank. However, zebrafish subjected to SCOP treatment showed an altered locomotor tracking style that was reversed by ELE administration; $p < 0.005$ for 3 mg/L and $p < 0.05$ for both 1 and 6 mg/L doses. In addition, a non-significant improvement in locomotion was observed after GAL treatment (Figure 13).

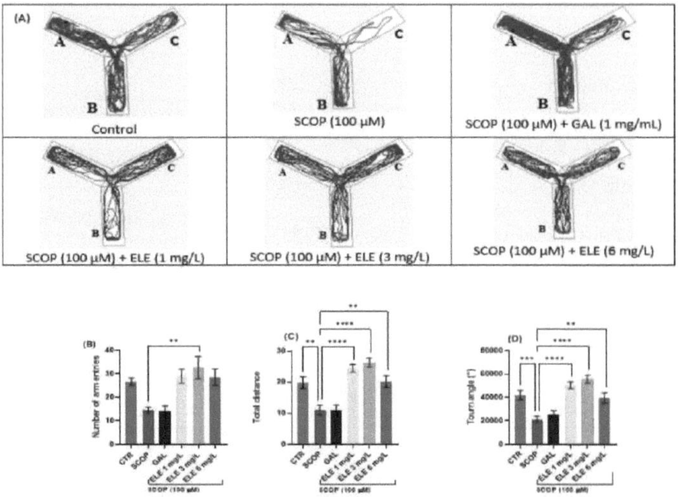

Figure 13. The effects of *Tribulus terrestris* (ELE, 1, 3 and 6 mg/L) administration in scopolamine (SCOP)-treated zebrafish on spatial memory and locomotor activity assessed in the Y-maze task. (A) The locomotor tracking pattern exhibited by zebrafish during the second phase of the Y-maze task as a function of their respective allocations to experimental

groups. Maze arms were designated A (starting arm), B (other arm) and C (new arm). (B) Number of entries in the arms. (C) Total distance (m). (D) Angle of rotation (°). Data are represented as means ± S.E.M. *(n = 8)*. ** $p < 0.001$, *** $p < 0.0001$ and **** $p < 0.00001$ (Tukey post hoc analyses). Galantamine (GAL, 1 mg/L) was used as a positive reference drug.

4. Effect of Tt extract on spatial memory using the Y maze test

ELE-treated fish explored all three arms of the maze. These fish outperformed the GAL-treated group in terms of behavioral parameters.

The Y maze was used to assess spatial working memory in amnesic fish. Figure 13 illustrates the memory impairment caused by SCOP, as SCOP-treated fish spent minimal time in the novel arm. ELE administration demonstrated memory-enhancing effects in amnesic fish. The one-way ANOVA test revealed a significant increase in spatial memory performance in the groups treated with low and high doses (1, 3 and 6 mg/L) of Tt in terms of total distance traveled ($F(8, 63) = 11.33$, $p < 0.0001$ for 1 and 3 mg/L and $p < 0.005$ for 6 mg/L), time spent in the novel arm (% of total time spent in the arms $F(8, 63) = 5.225$, $p < 0.01$) and turn angle (one-way ANOVA $F(8, 63) = 11.31$, $p < 0.001$), with increased percentages compared with the SCOP-only group, indicating significant effects on short-term memory. Further analysis using Tukey's post hoc test revealed a significant difference between the control and SCOP groups ($p < 0.0001$), control and SCOP + ELE (1 mg/L) ($p < 0.0001$), SCOP and SCOP + ELE (3 mg/L) ($p < 0.0001$), and SCOP and SCOP + ELE (6 mg/L) ($p < 0.001$), indicating that ELE significantly improved spatial memory (Figure 13). This improvement in spatial memory is consistent with findings in previous work investigating phytochemical effects and the Y-maze test (Brinza et al., 2020, 2022;

Capituna et al., 2020).

5. Effects of Tt extract on recognition memory using the NOR test

In this test, the object distinction task was performed to assess recognition memory in zebrafish. These vertebrates possess the ability to differentiate between familiar (OF) and unfamiliar objects and to remember three-dimensional geometric shapes (Dos Santos et al., 2018). Following the object discrimination test, we generated representative tracking plots illustrating the paths taken by the fish during the test session, according to their respective experimental groups. The effects of SCOP administration (100 μM) and ELE treatments (1, 3 and 6 mg/L) on NOR recognition memory and locomotion are illustrated in Figure 14. In the NOR test, the one-factor ANOVA test revealed a significant effect of treatment on preference percentages ($F (8, 61) = 17.30$, $p < 0.0001$) and exploration time ($F (17, 125) = 18.64$, $p < 0.0001$)) (Figure 14B). Representative locomotor tracking plots revealed changes in swimming behavior (Figure 14A). Compared with the control group, SCOP-treated fish showed lower percentages of preference for novel objects (NO), implying impaired recognition memory. Furthermore, ELE treatment (3 mg/L) increased preference percentages in SCOP-treated fish ($p < 0.05$ (Figure 14), suggesting memory enhancement. As shown in Figure 14, SCOP-treated fish spent a fair amount of time exploring OF and less time exploring NO compared to fish in the other groups. One-way ANOVA indicated statistical significance ($p < 0.05$). According to Figure 3, all three doses of ELE, particularly the 3 mg/L dose, reversed the recognition memory deficits caused by SCOP administration by significantly ($p < 0.05$ and $p < 0.01$ for ELE (3 mg/L)) increasing zebrafish preference for NO. The 3 mg/L doses of ELE demonstrated performance levels exceeding the chance level (50%), indicating a relative preference for NO. Therefore, ELE extract improves recognition memory in zebrafish with AD-induced SCOP patterns, as observed in the object distinction task. Saponins and flavonoids have been reported to improve recognition memory in laboratory animals (Kassem et al., 2020). In a previous study, flavonoids: rutin and naringin, improved discriminative ability in SCOP-treated animal models in a dose-dependent manner (Ramalingayya et al., 2016). The current results can be attributed to the abundance of these compounds in the Tt extract used.

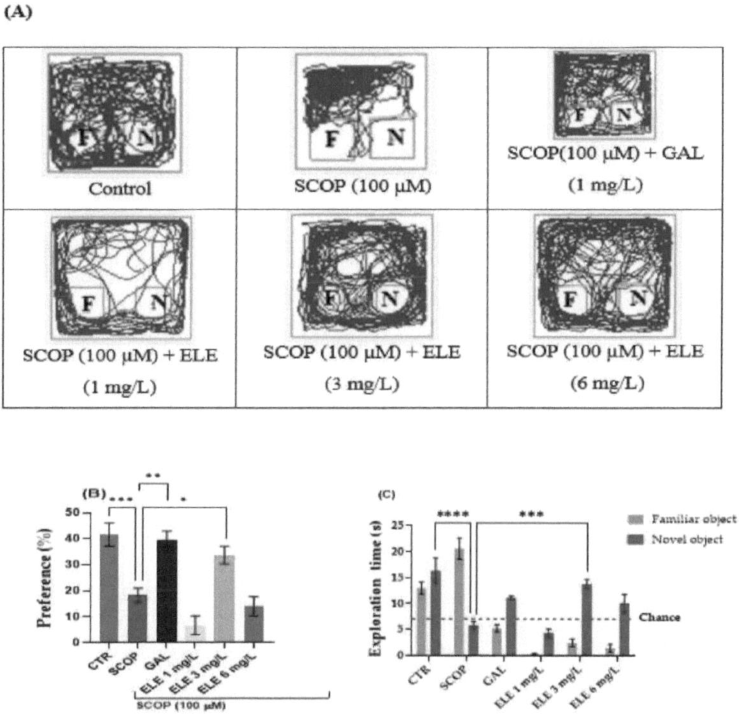

Figure 14. The effects of *Tribulus terrestris* (ELE, 1, 3 and 6 mg/L) administration to scopolamine (SCOP)-treated zebrafish on recognition memory, assessed through the object discrimination task. (A) Locomotor

tracking patterns of zebrafish during the test session of the object discrimination task according to their respective experimental groups. (B) Time spent on top (S). (C) The percentage preference measure was used as a final criterion to assess recognition memory. Familiar and novel objects were designated F and N, respectively. The black dashed line (chance) indicates a preference of 50%. Data are represented as means ± S.E.M. (n = 8). ** $p < 0.001$, *** $p < 0.0001$ and **** $p < 0.00001$ (Tukey post hoc analyses). Galantamine (GAL, 1 mg/L) was used as a positive reference drug.

Effects of Tt extract on brain AChE activity

To investigate the efficacy of Tt extract treatment on SCOP-treated zebrafish, key biochemical markers associated with cholinergic activity were measured. SCOP administration induced a significant increase in AChE activity ($p < 0.001$). This increase in AChE activity was significantly reduced in the GAL group to a level comparable to that of the control group ($p < 0.001$). Similarly, Tt treatment reduced AChE activity in a dose-dependent manner ($p < 0.005$ for 3 mg/L and $p < 0.01$ for 1 mg/L), compared with the groups treated with SCOP alone. The effects exerted by Tt exceeded those of GAL treatment (Figure 15A). These results are consistent with the inhibitory effects of Tt against AChE in animal models of neurodegenerative diseases (Saleem et al., 2020; Chauhdary et al., 2019).

Effects of Tt extract on cerebral oxidative status

One-factor analysis of variance (ANOVA) indicated significant treatment effects on brain oxidative status, as evidenced by changes in AChE ($F (8, 18) = 6.262, p < 0.0006$) (Figure 15A), SOD ($F (8, 18) = 9.002, p < 0.0001$) (Figure 15B), CAT ($F (8, 18) = 15.2, p < 0.0001$) (Figure 15C), GPX ($F (8, 18) = 22.82, p < 0.0001$) (Figure 15D), GSH ($F (8, 18) = 10.60, p < 0.0001$) (Figure 15E). High levels of lipid peroxidation (MDA) ($F (8, 18) = 16.02$ $p < 0.0001$) (Figure 15F) and protein oxidation (carbonylated proteins) ($F (8, 18) = 10, p < 0.0001$) were also observed (Figure 15G).

Compared to the control group, zebrafish exposed to SCOP showed a significant decrease in the activities of antioxidant enzymes such as superoxide dismutase (SOD) ($p < 0.0001$) (Figure 5B), GPX ($p < 0.0001$) (Figure 4D) and CAT ($p < 0.0001$) (Figure 4C), as well as elevated levels of

MDA ($p < 0.0001$) (Figure 4F) and carbonylated proteins ($p < 0.0001$) (Figure 4G). ELE treatment increased antioxidant enzyme activity by reducing SCOP-induced cerebral oxidative stress: SOD ($p < 0.005$ for 1 mg/L, 3 mg/L and $p < 0.001$ for 6 mg/L) (Figure 4), GPX ($p < 0.0001$ for 1 mg/L, 3 mg/L and 6 mg/L), CAT ($p < 0.005$ for 1 mg/L, 3 mg/L and $p < 0.001$ for 6 mg/L), GSH ($p < 0.001$ for 6 mg/L), as well as a reduction in MDA ($p < 0.0001$ for 1 mg/L, 3 mg/L and 6 mg/L) (Figure 4F) and carbonylated protein levels ($p < 0.001$ for 1 mg/L, 3 mg/L and 6 mg/L) (Figure) compared with SCOP-treated zebrafish.

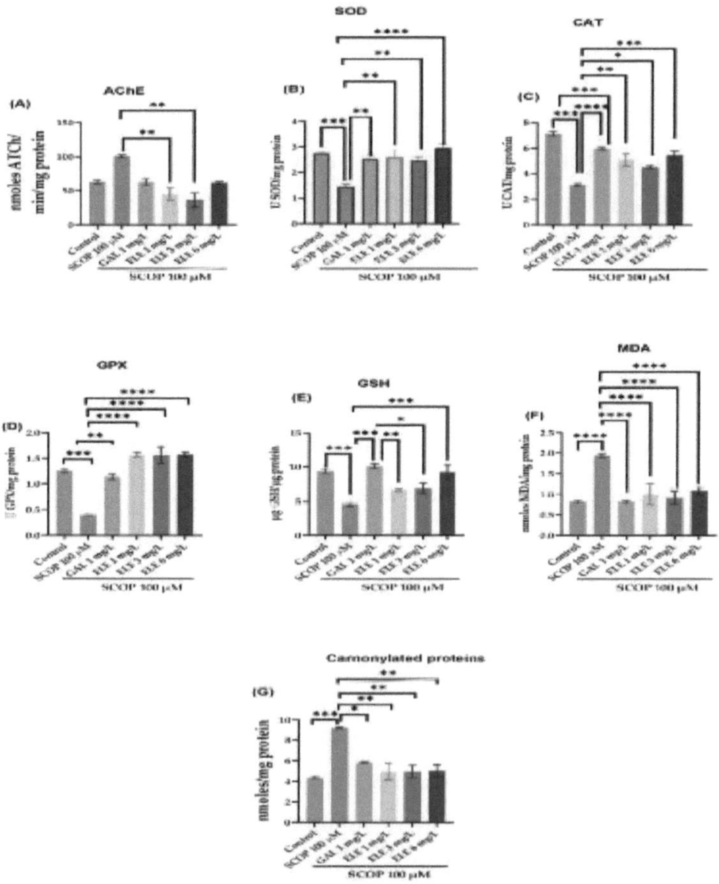

Figure15. The effects of *Tribulus terrestris* (ELE, 1, 3 and 6 mg/L) administration in scopolamine (SCOP)-treated zebrafish on (A) AChE; (B) superoxide dismutase (SOD) ; (C) catalase-specific activities (CAT); (D) glutathione peroxidase (GPX); (E) GSH; (F) malondialdehyde (MDA); (G) carbonylated protein levels. Data are represented as means ± S.E.M. *(n* = 8). ** $p < 0.001$, *** $p < 0.0001$ and **** $p < 0.00001$ (Tukey post hoc analyses). Galantamine (GAL, 1 mg/L) was used as a positive reference drug.

Discussion

Several studies have shown the presence of flavonoids in Tt, mainly quercetin, kaempferol and isorhamnetin derivatives (Zhu et al., 2017). In line with other published work, compounds similar to those presented here have been identified in Tt leaves (Stefanescu et al., 2020; Zhu et al.,2017). However, most published data provide general identifications without quantification of these compounds.

Extractive power correlates with plant material, solvent, geographical locality and preparation process (Sarvin et al., 2018; Zhu et al., 2017). The chemical composition shows that the extract studied is rich in flavonoids and saponins, which are known natural antioxidants. Their free hydroxyl groups enable interactions with cell membranes and trigger protective actions against free radicals. In addition, rutin, quercetin and catechin derivatives act as barriers against reactive oxygen species and hinder oxidative enzymes such as lipoxygenase, preventing cell denaturation (Alam et al., 2017).

Using molecular docking, the interactions of the 15 compounds (Tables 2) with AChE were visualized. The co-crystallized ligand donepezil was re-docked into the active site of AChE (Figure 16A) to assess docking accuracy. The root-mean-square deviation (RMSD) between the docking poses and the original co-crystallized poses was 0.47 Â.

The majority of selected compounds showed favorable interactions with essential amino acid residues such as Tyr72, Asp74, Trp86, Tyr124, Ser293, Phe295, Phe297, Tyr337, Phe338, Tyr341 and Trp286. Interestingly, rutin, with a docking score of -24.68 kcal/mol, showed 11 conventional hydrogen bonding interactions with essential amino acids from the PAS and CAS regions: Tyr124 (1.64 Â), Phe295 (2.93 Â), Tyr341 (2.45 Â), Tyr341 (3.04 Â), His447 (2.85 Â), Asp74 (2.71 Â), Thr83 (2.79 Â),

Asn87 (2.50 Â), Trp86 (3.12 Â), and Glu202 (3.05 Â and 3.22 Â).

In addition, rutin forms six carbon-hydrogen bonds, two π-donor hydrogen bonds, and eight π-π stacked bonds (Figure 16B, Table 3). Protodioscin exhibited a similar binding mode to rutin and donepezil (co-crystal ligand). It formed 16 hydrogen bonds with 12 amino acids, with lengths ranging from 2.04 to 3.23 Â.

The glycosyl group plays a key role in the formation of hydrogen bonds with Trp286 (3.23 Â), Trp86 (3.20 Â), Tyr72 (2.97 Â), Asp74 (2.82 and 2,65 Â), Tyr124 (2.04 Â), Tyr341 (2.04 and 2.81 Â), Ser293 (2.15 and 2.60 Â), Asn87 (2.80 and 2.70 Â), Thr83 (2.57 and 2.78 Â), Glu202 (2.77 Â), and Glu292 (2.47 Â).

The steroid moiety of protodioscin contributed to 14 hydrophobic interactions through the formation of π-Sigma or π-Alkyl interactions (Figure 16C, Table 3). In addition, saponin C (Figure 16D, Table 3) and terrestrosin C (Figure 16E, Table 3) exhibited extremely low binding energies of -24.11 and -24.63 Kcal/mol, respectively. Their steroid fractions exhibited hydrophobic interactions with Try72, Try286, Phe297, Phe338 and Try341 in the PAS region of the AChE enzyme. In addition, saponin C and terrestrosin C formed twelve and twenty-one hydrogen bonds respectively.

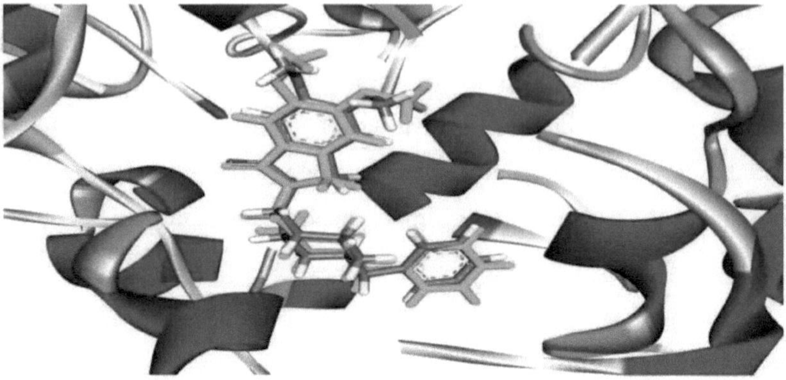

(A)

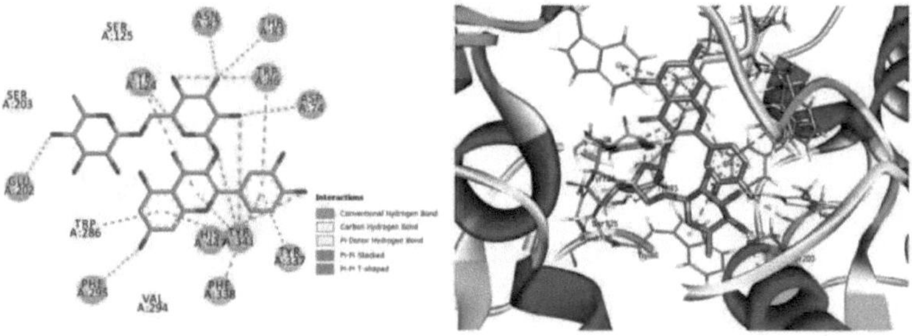
SER
A:125
SER
A:203
TRP
A:286
VAL
A:294
Interactions
Conventional Hydrogen Bond
Carbon Hydrogen Bond
Pi-Donor Hydrogen Bond
Pi-Pi Stacked
Pi-Pi T-shaped

(B)

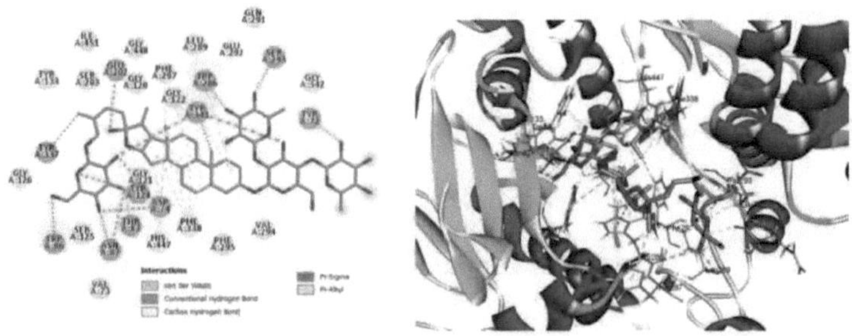
Interactions
van der Waals
Conventional Hydrogen Bond
Carbon Hydrogen Bond
Pi-Sigma
Pi-Alkyl

(C)

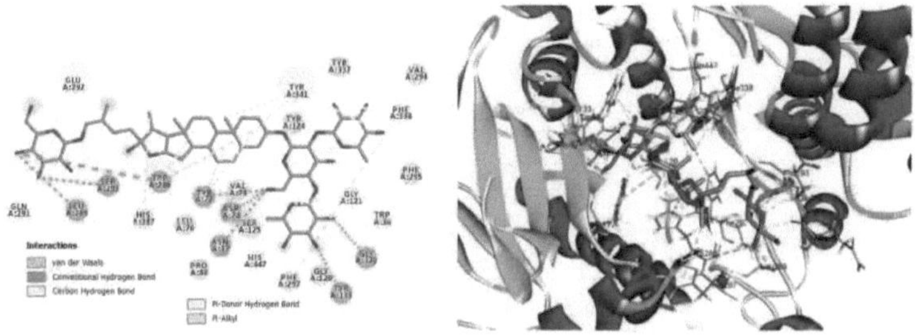

(D)

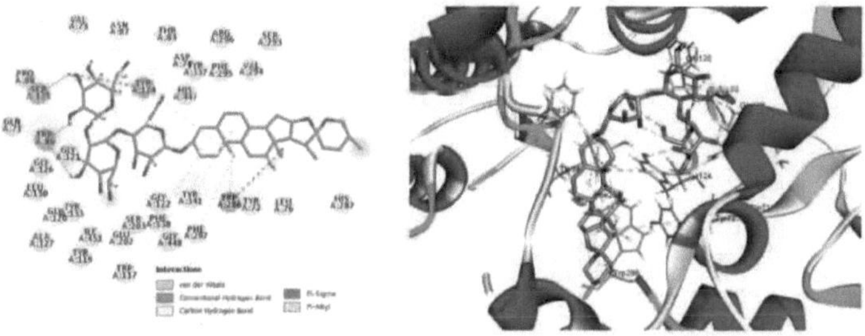

(E)

Figure 16: (A) Superposition of the co-crystalline ligand (donepezil) (in blue) with its docked pose (in grey) showed an appreciable RMSD of 0.47 Â. (B) The 2D (left) and 3D (right) binding mode of rutin in the active site of the AChE enzyme. (C) The 2D (left) and 3D (right) binding mode of protodioscin in the active site of the AChE enzyme. (D) 2D (left) and 3D (right) binding of saponin C in the AChE active site. (E) 2D (left) and 3D (right) binding of terrestrosine C in the AChE active site.

Table3. Interactions of Donepezil, Terrestrosin C, Protodioscin, Rutin and Saponin C in the active site of the AChE enzyme.

Affinity energy (kcal/mol)	Interaction	Amino Acid (Distance in angstrom (Å)

Donepezil -14.62 **(Co-crystal ligand)**	Attractive Load	Asp74 (5.40)
	[1]Hydrogen Bond	Phe295 (1.96)
	[2]Hydrogen Bond	Ser293 (3.06), Tyr72 (3.20)
	π -Cation	Trp86 (4.73), Tyr337 (3.91)
	π -Sigma	Tyr341 (3.59), Trp286 (3.64)
	π - π Stacked	Trp86 (4.46), Trp86 (3.89, 3.82,5.11), Tyr341 (5.05)
Terrestrosin -24.11 **C**	[2]Hydrogen Bond	Gly120 (2.93, 2.88), Asp74 (2.37), Glu202 (2.27), Asn87 (2.33, 2.34), Tyr124 (2.43), Trp86 (2.00)
	[1]Hydrogen Bond	Tyr124 (2.13), Ser125 (1.76), Trp86 (2.52, 1.68)
	π -Alkyl	Tyr72 (4.65), Tyr124 (4.26), Trp286 (3.20, 4.21, 4.48), Phe297 (4.98), Phe338 (4.91), Tyr341 (4.59, 3.67)
	π -Sigma	Trp286 (2.63)
Protodioscin -24.58	[1]Hydrogen Bond	Tyr72 (2.97), Asp74 (2.82), Tyr124 (2.04), Ser293 (2.15, 2.60), Tyr341 (2.75, 2.81), Trp286 (3.23), Trp86 (3.20), Asn87 (2.80, 2.70), Thr83 (2.57), Asp74 (2.65), Glu202 (2.77)
	[2]Hydrogen Bond	Thr83 (2.78), Glu292 (2.47)
	π -Sigma	Tyr341 (3.48), Tyr337 (3.00)

		π -Alkyl	Tyr124 (4.94), Trp286 (4.76), Phe297 (5.18, 4.56), Phe338 (5.13, 5.01, 4.31, 5.49), Tyr341 (3.80, 4.53), His447 (4.69, 5.41)
Rutin	-24.68	1Hydrogen Bond	Tyr124 (1.64), Phe295 (2.93), Tyr341 (2.45, 3.04), His447 (2.85), Asp74 (2.71), Thr83 (2.79), Asn87 (2.50), Trp86 (3.12), Glu202 (3.05, 3.22)
		2Hydrogen Bond	Trp86 (2.83), Asn87 (2.74), Ser125 (3.02), Ser203 (3.03), Val294 (2.41), His447 (2.36)
		3Hydrogen Bond	His447 (2.91), Trp286 (3.54)
		π - π Stacked	Trp286 (4.85), Tyr337 (3.44), Tyr341 (3.32, 4.03, 5.61)
		π - π T-shaped	Trp86 (5.92), Tyr124 (5.88), Phe338 (4.80)
Saponin C	-24.63	1Hydrogen Bond	Asp74 (2.50), Tyr133 (2.96), Ser293 (2.70), Tyr72 (2.90), Asn87 (3.23), Trp286 (3.36), Leu289 (2.99), Gly120 (2.88)
		2Hydrogen Bond	Gly121 (2.60, 2.69, 2.16), Glu292 (2.84), Phe338 (2.38, 2.41), His447 (2.98), Gln291 (3.54)
		3Hydrogen Bond	Tyr341 (3.58), Trp86 (3.62, 4.12), Phe338 (3.02, 3.73)
		π-Alkyl	Tyr72 (5.33, 4.87), Trp286 (5.33, 4.63, 5.26, 3.88), His287 (4.67), Phe297 (5.29), Phe338 (5.27), Tyr341 (4.84, 4.88)

1 Conventional hydrogen bond. 2 Carbon hydrogen bond. 3 π-donor hydrogen bond

The inhibitory properties of Tt on AChE may be due to the presence of various flavonoids and saponins, including epigallocatechin, apigetrin, rutin, quercetin, luteolin, cynaroside, caffeic acid, trillin, hecogenin, terreside B, trillarin, protodioscin and saponin C (Figure 1).

Moreover, during docking, these compounds showed good interactions with crucial amino acid residues such as Tyr72, Asp74, Trp86, Tyr124, Ser293, Phe295, Phe297, Tyr337, Phe338, Tyr341 and Trp286. The most stable docked molecules were rutin, saponin C, terrestrosin C and protodioscin, all with docking energies below -24 kcal/mol.

Interestingly, the maximum docking score for rutin was achieved at -24.68 kcal/mol. This suggests that rutin plays a significant role as an AChE inhibitor. A recent study by Sleem et al, examined the antiparkinsonian potential of Tt using a haloperidol-induced pathological disorder model in rats (Saleem et al.,2020). This study showed that the plant's methanolic extract possesses antioxidant potential, promoting disease amelioration by increasing levels of endogenous antioxidant enzymes and reducing levels of AChE, a-synuclein, IL-1β and TNF-α (Jaworski et al., 2019; Yang et al.,2017)

Other studies have indicated that triterpene pentacyclic saponins similar to astragaloside can protect the survival of PC12 neuronal cells by activating GSK-3β and preventing the opening of the mitochondrial permeability transition pore (mPTP).

GSK-3β plays an important role in neuronal cell survival, growth and neuronal plasticity (Jaworski et al., 2019; Fu et al., 2020). Activation of GSK-3β to prevent mPTP opening is essential for neurological protection (Fu et al., 2020; Yamazaki et al., 2019).

Ginsenosides, gintonin, ginseng extracts/fractions and ginseng-containing formulas have been extensively studied in cells, demonstrating that ginseng possesses neuroprotective properties against Alzheimer's disease. These effects are exerted via the regulation of several signaling pathways, such as the PI3K/Akt, AMPK-mTOR and NFκB pathways, to inhibit or ameliorate key features of AD, including β-amyloid accumulation, neurotrophic factors, tau protein phosphorylation, neuroinflammation, apoptosis and mitochondrial dysfunction at different stages of AD (Li et al., 2019).

Blockade of the central muscarinic acetylcholine receptor leads to impaired learning and memory in humans and animals. The anticholinergic SCOP has been used as a potent amnesic agent. Various studies have demonstrated that prolonged administration of SCOP leads to a reduction in protective antioxidant mechanisms. This occurs due to suppression of nuclear factor erythroid 2 (Nrf2)-related factor 2 expression and is linked to reduced synaptic plasticity via the CREB/BDNF (cAMP response element-binding protein/brain-derived neurotrophic factor) pathway (Ransey et al., 2017; Yamazaki et al., 2019).

Furthermore, SCOP-induced oxidative stress phosphorylates GSK and increases AChE activity, ultimately leading to apoptosis via Bax/Bcl-2 mediators (Ransey et al., 2017).

In AD patients, ACh synthesis and release are delayed, while AChE is hyperactive and further degrades ACh, obstructing the transmission of the cholinergic system (Cao et al., 2018; Sultana et al., 2013).

Aged animal models have revealed a significant association between the development of behavioral deficits or cognitive impairment in learning and memory, and spatial and temporal memory (Sultana et al., 2013).

In the SCOP-induced model, Tt inhibited AChE activity, demonstrating positive effects on cholinergic neurotransmission. Carbonylated proteins are the products of protein oxidation (Cao et al., 2018), while CAT and SOD are considered two important antioxidant enzymes that combat oxidative stress (Concalves et al., 2020; Song et al., 20121; Yin et al., 2020).

Conclusion

The results of this study suggest that administration of ethanolic extract of *Tribulus terrestris* leaves improves cognitive dysfunction and memory impairment, as measured by the Y-maze and NOR behavioral tasks. In addition, Tt treatment reversed SCOP-induced oxidative stress by increasing antioxidant enzyme activities and reducing carbonylated protein and MDA levels.

Our results also indicate significant inhibition of AChE activity after treatment, a reliable approach in the management of neurodegenerative diseases, and this inhibition is also corroborated by in silico data.

These results highlight the memory-enhancing effects of Tt extracts and their potential as promising candidates for the management of Alzheimer's disease.

References

1. **Ablain J**, Zon LI. Of fish and men: using zebrafish to fight human diseases. Trends Cell Biol. 2013; 23:584-586. [PubMed: 24275383]
2. **Ahmad F**, et al. Zebrafish embryos and larvae in behavioural assays. Behavior. 2012; 149:1241- 1281.
3. **Ahmadi. A** and Shadboorestan A., "Oxidative stress and cancer; the role of hesperidin, a *citrus* natural bioflavonoid, as a cancer Chemoprotective agent," Nutrition and Cancer, vol. 68, no. 1, pp. 29-39, 2016.
4. **AL-ALI M. et al,** *Tribulus terrestris:* preliminary study of its diuretic and contractile effects and comparison with Zea Mays. *JEthnopharmacology.* **2003**, 85: 257-60.
5. **Alam, P.;** Parvez, M.K.; Arbab, A.H.; Al-Dosari, M.S. Quantitative analysis of rutin, quercetin, naringenin, and gallic acid by validated RP- and NP-HPTLC methods for quality control of anti-HBV active extract of Guiera senegalensis. Pharm. Biol. 2017, 55, 1317-1323
6. **Alzheimer's Association**. Available from: // www.alz.org /en/ quest-ce-que-la- maladie-dalzheimer.asp. [cited 26 Jul 2021].
7. **Ara, A**.; Vishvkarma, R.; Mehta, P.; Rajender, S., The Profertility and Aphrodisiac Activities of *Tribulus terrestris L.*: Evidence from Meta-Analyses. *Andrologia* **2023,** 2023, 7118431.
8. **Benninghoff, J**.; Perneczky, R., Anti-Dementia Medications and AntiAlzheimer's Disease Drugs: Side Effects, Contraindications, and Interactions. In *Neuro Psychopharmacotherapy,* Riederer, P.; Laux, G.; Nagatsu, T.; Le, W.; Riederer, C., Eds. Springer International Publishing: Cham, 2020; pp 1-10.
9. **Bonnet H.** Contribution a l'étude de *Tribulus Terrestris L.* Thèse de pharmacie. Université de Lorraine. Faculté de pharmacie de Nancy.

2003.

10. **Bonnier G**. Flore complète de France Suisse et Belgique de Gaston Bonnier, Tome II. Librairie générale de l'enseignement, **1911**, p104-110.

11. **Bouabdallah S,** Sgheier RM, Salmi S, Khalifi D, Laouni D, B-Attia M. Current Approachcs and Challenges for Chemical Characterization of inhibitory effect against cancer Cell line isolated from Gokshur Extract. *J Chromatography B,* 2016; 1026(C): 279-285.

12. **Bouabdallah, S**.; Bouzouit, N.; Laouini, D.; El Bok, S.; Sghaier, R.; Selmi, S.; Attia, M., Separation and evaluation of natural antileishmanial potential against leishmania major and infantum isolated from the Tunisia strains. **2018**.

13. **Bouabdallah, S.**; Al-Maktoum, A.; Amin, A., Steroidal Saponins: Naturally Occurring Compounds as Inhibitors of the Hallmarks of Cancer. *Cancers (Basel)* **2023,** 15, (15).

14. **Buske C**, Gerlai R. Early embryonic ethanol exposure impairs shoaling and the dopaminergic and serotoninergic systems in adult zebrafish. Neurotoxicol. Teratol. 2011; 33:698-707. [PubMed: 21658445]

15. **Buske C,** Gerlai R. Shoaling develops with age in zebrafish (Danio rerio). Prog. Neuropsychopharmacol. Biol. Psychiatry. 2011; 35:1409-1415. [PubMed: 20837077]

16. **Brennan CH**. Zebrafish behavioural assays of translational relevance for the study of psychiatric disease. Rev. Neurosci. 2011; 22:37-48. [PubMed:

21615260]

17.**Brinza, I.;** Abd-Alkhalek, A. M.; El-Raey, M. A.; Boiangiu, R. S.; Eldahshan, O. A.; Hritcu, L., Ameliorative Effects of Rhoifolin in Scopolamine-Induced Amnesic Zebrafish (Danio rerio) Model. *Antioxidants* **2020,** 9, (7), 580.

18.**Brinza, I.;** Raey, M. A. E.; El-Kashak, W.; Eldahshan, O. A.; Hritcu, L., Sweroside Ameliorated Memory Deficits in Scopolamine-Induced Zebrafish (Danio rerio) Model: Involvement of Cholinergic System and Brain Oxidative Stress. *Molecules* **2022,** 27, (18), 5901.

19.**Cachat J,** et al. Measuring behavioral and endocrine responses to novelty stress in adult zebrafish. Nat. Protoc. 2010; 5:1786-1799. [PubMed: 21030954].

20. **Cachat J**, et al. Unique and potent effects of acute ibogaine on zebrafish: the developing utility of novel aquatic models for hallucinogenic drug research. Behav. Brain Res. 2013; 236:258-269. [PubMed: 22974549]

21. **Cai, B**.; Zhang, Y.; Wang, Z.; Xu, D.; Jia, Y.; Guan, Y.; Liao, A.; Liu, G.; Chun, C.; Li, J., Therapeutic Potential of Diosgenin and Its Major Derivatives against Neurological Diseases: Recent Advances. *Oxidative Medicine and Cellular Longevity* **2020,** 2020, 3153082.

22.5. Grossman L, et al. Characterization of behavioral and endocrine effects of LSD on zebrafish. Behav. Brain Res. 2010; 214:277-284. [PubMed: 20561961]

23. **Cao, P.;** Sun, J.; Sullivan, M.A.; Huang, X.; Wang, H.; Zhang, Y.; Wang, N.; Wang, K. Angelica sinensis polysaccharide protects against acetaminopheninduced acute liver injury and cell death by suppressing oxidative stress and hepatic apoptosis in vivo and in vitro. Int. J. Biol. Macromol. 2018, 111, 11331139.

24. **Capatina, L.;** Todirascu-Ciornea, E.; Napoli, E. M.; Ruberto, G.; Hritcu, L.; Dumitru, G., Thymus vulgaris Essential Oil Protects Zebrafish against Cognitive Dysfunction by Regulating Cholinergic and Antioxidants Systems. *Antioxidants* **2020,** 9, (11), 1083.
25. **Chen, W. N.**; Yeong, K. Y., Scopolamine, a Toxin-Induced Experimental Model, Used for Research in Alzheimer's Disease. *CNS Neurol Disord Drug Targets* **2020,** 19, (2), 85-93.
26. **Chhatre, S.;** Nesari, T.; Somani, G.; Kanchan, D.; Sathaye, S., Phytopharmacological overview of Tribulus terrestris. *Pharmacogn Rev* **2014,** 8, (15), 45-51.
27. **Canu N**, Amadoro G, Triaca V, Latina V, Sposato V, Corsetti V, et al. The Intersection of NGF/TrkA Signaling and Amyloid Precursor Protein Processing in Alzheimer's Disease Neuropathology. Int J Mol Sci. June 20, 2017;18(6).
28. **Chadefaud**. Traité de botanique systématique, Tome II, Fascicule II. Masson, 1960p 863.
29. **Collier AD**, Echevarria DJ. The utility of the zebrafish model in conditioned place preference to assess the rewarding effects of drugs. Behav. Pharmacol. 2013; 24:375-383. [PubMed: 23811781]
30. **Delmas C**. Communiquer avec la personne atteinte de maladie d'Alzheimer et troubles apparentés Sensibilisation à la méthodologie de soins Gineste Marescotti, Maintien à domicile, 6th year course, 2020.
31. **Dinchev** D, Janda B, Evstatieva L, Oleszek W, Aslani M.R, **KostovaI.** Distribution of steroidal saponins in *Tribulus terrstris* from differnt geographical regions. *Phytochemistry.* 2008; 69: 176-186.
32. **Duffner PM**. Alzheimer's disease: current etiological knowledge, future treatments, and various preventive means [Internet]. Available from: file:///C:/Users/Madem/Documents/School/Th%C3%A8se%20sur%20

Alzhei mer/Th%C3%A8se%20Duffner%20Pierre-Marie.pdf

33. **Dunn, N. R.**; Pearce, G. L.; Shakir, S. A., Adverse effects associated with the use of donepezil.

3 4.in general practice in England. *JPsychopharmacol* **2000,** 14, (4), 406-8.

35. **Drew RE**, et al. Brain transcriptome variation among behaviorally distinct strains of zebrafish (Danio rerio). BMC Genomics. 2012; 13:323. [PubMed: 22817472]

36. **Ekeanyanwu R. C.** and Njoku O. U. "Flavonoid-rich fraction of the *Monodora tenuifolia* seed extract attenuates behavioural alterations and oxidative damage in forced swim stressed rats," Chinese Journal of Natural Medicines, vol. 13, no. 3, pp. 183-191, 2015.

37. **Filipovic D**. Todorovié N. Bernardi R. E. and Gass P. "Oxidative and nitrosative stress pathways in the brain of socially isolated adult male rats demonstrating depressive- and anxiety-like symptoms," Brain Structure and Function, vol. 222, no. 1, pp. 1-20, 2017.

38. **Fondation A.** How was Alzheimer's disease discovered? [Internet]. Alzheimer Foundation. 2019 [cited 26 Jul 2021]. Available from: https://www.fondation-alzheimer.org/comment-la-maladie-dalzheimer-a-t- elleete-decouverte/.

39. **VA Foundation.** Alzheimer's figures [Internet]. Fondation Vaincre Alzheimer. [cited 26 Jul 2021]. Available from: https://www.vaincrealzheimer.org/la-maladie/quelques-chiffres/

40. **Fu, Y.;** Cai, J.; Xi, M.; He, Y.; Zhao, Y.; Zheng, Y.; Zhang, Y.; Xi, J.; He, Y. Neuroprotection Effect of Astragaloside IV from 2-DG Induced Endoplasmic Reticulum Stress. Oxidative Med. Cell. Longev. 2020, 2020, 9782062. 51.

41. **Gerlai R**. Zebrafish antipredatory responses: a future for translational research? Behav. Brain Res. 2010; 207:223-231. [PubMed: 19836422]

42. **Gerlai R**. Using zebrafish to unravel the genetics of complex brain disorders. Curr. Top. Behav. Neurosci. 2012; 12:3-24. [PubMed: 22250005].

43. **Gonçalves, R.L.G**.; Cunha, F.V.M.; Sousa-Neto, B.P.S.; et al. α-Phellandrene attenuates tissue damage, oxidative stress, and TNF-α levels on acute model ifosfamide-induced hemorrhagic cystitis in mice. Naunyn Schmiedebergs Arch. Pharmacol. 2020, 393, 1835-1848.

44. **Goozee, K. G**.; Shah, T. M.; Sohrabi, H. R.; Rainey-Smith, S. R.; Brown, B.; Verdile, G.; Martins, R. N., Examining the potential clinical value of curcumin in the prevention and diagnosis of Alzheimer's disease. *British Journal of Nutrition* **2016,** 115, (3), 449-465.

45. **Griffiths BB,** et al. A zebrafish model of glucocorticoid resistance shows serotonergic modulation of the stress response. Front. Behav. Neurosci. 2012; 6:68. [PubMed: 23087630]

46. **Guan L. P.** and Liu B. Y., "Antidepressant-like effects and mechanisms of flavonoids and related analogues," European Journal of Medicinal Chemistry, vol. 121, pp. 47-57, 2016.

47. Guan, L.; Mao, Z.; Yang, S.; Wu, G.; Chen, Y.; Yin, L.; Qi, Y.; Han, L.; Xu, L., Dioscin alleviates Alzheimer's disease through regulating RAGE/NOX4 mediated oxidative stress and inflammation. *Biomedicine & Pharmacotherapy* **2022,** 152, 113248.

48. **Herrera-Ruiz M**, Zamilpa A, González-Cortazar M et al, "Antidepressant effect and pharmacological evaluation of standardized extract of flavonoids from Byrsonima crassifolia," Phytomedicine, vol. 18, no. 14, pp. 1255-1261, 2011.

4 9.Inserm. Alzheimer · Inserm, Science for health [Internet]. Inserm. [cited 26 Jul 2021]. Available from: https://www.inserm.fr/dossier/alzheimer-disease/

5 0.**Ivan A and Ross R**. Medicinal plants of the world chemical constituents, traditional and modern medicinal uses. *Humana Press,* Totowa: 2001; 2: 411426.

51.**Jaworski, T**.; Banach-Kasper, E.; Gralec, K. GSK-3β at the Intersection of Neuronal Plasticity and Neurodegeneration. Neural Plast. 2019, 2019, 4209475.

52.**Johansson, M**.; Stomrud, E.; Lindberg, O.; Westman, E.; Johansson, P. M.; van Westen, D.; Mattsson, N.; Hansson, O., Apathy and anxiety are early markers of Alzheimer's disease. *Neurobiol Aging* **2020,** 85, 74-82.

53.**Kalueff AV**, et al. Towards a comprehensive catalog of zebrafish behavior 1.0 and beyond. Zebrafish. 2013; 10:70-86. [PubMed: 23590400]

54.**Kalueff,** AV.; Stewart, A. M.; Gerlai, R., Zebrafish as an emerging model for studying complex brain disorders. *Trends Pharmacol Sci* **2014,** 35, (2), 63-75.

55.**Kalueff AV,** et al. Zebrafish as an emerging model for studying complex brain disorders. Trends Pharmacol. Sci. 2014; 35:63-75. [PubMed: 24412421]

56.**Kalueff, AV.,** et al. Gaining translational momentum: more zebrafish models for neuroscience research. Prog. Neuropsychopharmacol. Biol. Psychiatry. 2014. http://dx.doi.org/10.1016/j.pnpbp. 2014.01.022

57.**Khalid, A**.; Algarni, A. S.; Homeida, H. E.; Sultana, S.; Javed, S. A.; Rehman, Z. u.; Abdalla, H.; Alhazmi, H. A.; Albratty, M.; Abdalla, A. N., Phytochemical, Cytotoxic, and Antimicrobial Evaluation of *Tribulus terrestris L.,* Typha domingensis Pers., and Ricinus communis L.: Scientific Evidences for Folkloric Uses. *Evidence-Based Complementary and Alternative Medicine* **2022,** 2022, 6519712.

58.**Khalid, A**.; Nadeem, T.; Khan, M. A.; Ali, Q.; Zubair, M., In vitro evaluation of immunomodulatory, anti-diabetic, and anti-cancer molecular mechanisms of *Tribulus terrestris* extracts. *Scientific Reports* **2022,** 12, (1), 22478.

59. **Khan, K.** M.; Collier, A. D.; Meshalkina, D. A.; Kysil, E. V.; Khatsko, S. L.; Kolesnikova, T.; Morzherin, Y. Y.; Warnick, J. E.; Kalueff, A. V.; Echevarria, D. J., Zebrafish models in neuropsychopharmacology and CNS drug discovery. *Br JPharmacol* **2017,** 174, (13), 1925-1944.

60. **Kanimozhi S.**, Bhavani P., and Subramanian P., "Influence of the flavonoid, quercetin on antioxidant status, lipid peroxidation and histopathological changes in hyperammonemic rats," Indian Journal of Clinical Biochemistry, vol. 32, no. 3, pp. 275-284, 2017.

61. **Kassem, I. A. A.**; Ragab, M. F.; Nabil, M.; Melek, F. R., Two new acylated triterpenoidal saponins from Gleditsia caspica Desf. and the effect of its saponin content on LPS-induced cognitive impairment in mice. *Phytochemistry Letters* **2020,** 40, 53-62.

62. **Kareti, S**. R.; P, S., In Silico Molecular Docking Analysis of Potential AntiAlzheimer's Compounds Present in Chloroform Extract of Carissa carandas Leaf Using Gas Chromatography MS/MS. *Current Therapeutic Research* **2020,** 93, 100615.

63. **Kim, J**.; Lee, H. J.; Lee, K. W., Naturally occurring phytochemicals for the prevention of Alzheimer's disease. *JNeurochem* **2010,** 112, (6), 1415-30.

64. **Kim HJ,** Kim JC, Min JS, Kim MJ, Kim JA, Kor MH, Yoo HS, Ahn JK. Aqueous extract of tribulus terrestris Linn induces cell growth arrest and apoptosis by down-regulating NF-kB signaling in liver cancer cells. *J Ethnopharmacol.* 2011; 136: 197- 203.

65. **Kyzar EJ**, et al. Effects of hallucinogenic agents mescaline and phencyclidine on zebrafish behavior and physiology. Prog.

Neuropsychopharmacol. Biol. Psychiatry. 2012; 37:194-202. [PubMed: 22251567]

66. **Levin, E.** D.; Bencan, Z.; Cerutti, D. T., Anxiolytic effects of nicotine in zebrafish. *Physiol Behav* **2007,** 90, (1), 54-8.

67. **Li, G.;** Zhang, N.; Geng, F.; Liu, G.; Liu, B.; Lei, X.; Li, G.; Chen, X., High-throughput metabolomics and ingenuity pathway approach reveals the pharmacological effect and targets of Ginsenoside Rg1 in Alzheimer's disease mice. *Scientific Reports* **2019,** 9, (1), 7040.

68. **Liao, Y.;** Mai, X.; Wu, X.; Hu, X.; Luo, X.; Zhang, G., Exploring the Inhibition of Quercetin on Acetylcholinesterase by Multispectroscopic and In Silico Approaches and Evaluation of Its Neuroprotective Effects on PC12 Cells. *Molecules* **2022,** 27, (22).

69. **Liu Z.**, Zhou T., Ziegler A.C., Dimitrion P., and L. Zuo, "Oxidative stress in neurodegenerative diseases: from molecular mechanisms to clinical applications," Oxidative Medicine and Cellular Longevity, vol. 2017, Article ID 2525967, 11 pages, 2017.

70. **Livingston G.**, Huntley J., Sommerlad A., et al, Dementia prevention, intervention, and care: 2020 report of the Lancet Commission. *Lancet* **2020,** 396, (10248), 413-446.

71. **López Patiño MA**, et al. Gender differences in zebrafish responses to cocaine withdrawal. Physiol. Behav. 2008; 95:36-47. [PubMed: 18499199]

72. **Lulita MF**, Bistué Millón MB, Pentz R, Aguilar LF, Do Carmo S, Allard S, et al. Differential deregulation of NGF and BDNF neurotrophins in a transgenic rat model of Alzheimer's disease. Neurobiol Dis. Dec 2017; 108:307-23.

73. **MSD manuals** for the general public. Confusional syndrome - Disorders of the brain, spinal cord and nerves [Internet]. [cited 2022 Aug 12].

Available from: https://www.msdmanuals.com/fr/accueil/troubles-du-cerveau,- de-lamoelle-% C3 % A9 pini%C3 % A 8re-et-des-nerfs/syndrome-confusionnel-et d% C3% A9 mence / syndrome-confusionnel

74. **Maurya P.K.**, C. Noto C., Rizzo L. B. et al, "The role of oxidative and nitrosative stress in accelerated aging and major depressive disorder," Progress in Neuro-Psychopharmacology and Biological Psychiatry, vol. 65, pp. 134-144, 2 016.
75. **Miller N**, et al. Effects of nicotine and alcohol on zebrafish (Danio rerio) shoaling. Behav. Brain Res. 2013; 240:192-196. [PubMed: 23219966].
76. **Moreira, N.;** Lima, J.; Marchiori, M. F.; Carvalho, I.; Sakamoto-Hojo, E. T., Neuroprotective Effects of Cholinesterase Inhibitors: Current Scenario in Therapies for Alzheimer's Disease and Future Perspectives. *J Alzheimers Dis Rep* **2022,** 6, (1), 177-193.
77. **Muller A**. Troubles psycho-comportementaux & cognitifs, La personne âgée, DFASP3 course, 2020.
78. **Neelkantan N,** et al. Perspectives on zebrafish models of hallucinogenic drugs and related psychotropic compounds. ACS Chem. Neurosci. 2013; 4:11371150. [PubMed: 23883191]
79. **Newman, M.** Using the zebrafish model for Alzheimer's disease research. *Front Genet* **2014,** 5, 189.
80. **Parker MO**, et al. Development and automation of a test of impulse control in zebrafish. Front. Syst. Neurosci. 2013; 7:65. [PubMed: 24133417].
81. **Parker MO,** et al. Discrimination reversal and attentional sets in zebrafish (Danio rerio). Behav. Brain Res. 2012; 232:264-268. [PubMed: 22561034]

82. **Parker MO,** et al. Discrimination reversal and attentional sets in zebrafish (Danio rerio). Behav. Brain Res. 2012; 232:264-268. [PubMed: 22561034]
83. **Payet O**. Memory, Physiology, DFASP1 course, 2018.
84. **Phillips, O. A.**; Mathew, K. T.; Oriowo, M. A., Antihypertensive and vasodilator effects of methanolic and aqueous extracts of Tribulus terrestris in rats. *Journal OfEthnopharmacology* **2006,** 104, (3), 351-355.
85. **Pippal JB**, et al. Characterization of the zebrafish (Danio rerio) mineralocorticoid receptor. Mol. Cell. Endocrinol. 2011; 332:58-66. [PubMed: 20932876]
86. **Poucheret P**. Medications & Alzheimer's, Pharmacology, DFASP1 course, 2018.
87. Fahnestock M, Shekari A. ProNGF and Neurodegeneration in Alzheimer's Disease. Front Neurosci. 2019; 13:129.
88. **Psychomédia**. DSM-5 : le GUIDE Psychomédia. [cited 18 Feb 2022]. Available from: http://www.psychomedia.qc.ca/dsm-5/2013-05-22/guide- psychomedia.
89. **Rahman, M. A**, Rahman, M. S, Uddin, M. J, Mamum-Or-Rashid, A. N. M, Pang, M.-G.; Rhim, H., Emerging risk of environmental factors: insight mechanisms of Alzheimer's diseases. *Environmental Science and Pollution Research* **2020,** 27, (36), 44659-44672.
90. **Ramalingayya, G. V.;** Nampoothiri, M.; Nayak, P. G.; Kishore, A.; Shenoy, R. R.; Mallikarjuna Rao, C.; Nandakumar, K., Naringin and Rutin Alleviate Episodic Memory Deficits in Two Differentially Challenged Object Recognition Tasks. *Pharmacogn Mag* **2016,** 12, (Suppl 1), S63-70.
91. **Ramsey, C.P.;** Glass, C.A.; Montgomery, M.B.; Lindl, K.A.; Ritson, G.P.; Chia, L.A.; Hamilton, R.L.; Chu, C.T.; Jordan-Sciutto, K.L.

Expression of Nrf2 in neurodegenerative diseases. J. Neuropathol. Exp. Neurol. 2007, 66, 75-85. 53.

92. **Saleem, S.**; Kannan, R. R., Zebrafish: an emerging real-time model system to study Alzheimer's disease and neurospecific drug discovery. *Cell Death Discov* **2018,** 4, 45.

93. **Saleem, U.**; Chauhdary, Z.; Raza, Z.; Shah, S.; Rahman, M.-u.; Zaib, P.; Ahmad, B. Anti-Parkinson's Activity of Tribulus terrestris via Modulation of AChE, α- Synuclein, TNF-α, and IL-1β. ACS Omega 2020, 5, 25216-25227.

94. **Salim S.** "Oxidative stress and the central nervous system," The Journal of Pharmacology and Experimental Therapeutics, vol. 360, no. 1, pp. 201-205,
2 017.

95. **Sharifi A. M. et al,** Study of antihypertensive mechanism of *Tribulus terrestris* in 2K1C hypertensive rats: Role of tissue ACE activity. *Life Sciences.* 2003; (73): 2963-71.

96. **Samuelsson G,** Farah MH, Cleason P, Hagos M, Thulin M, Hedberg O, Warfa AM, hassan AO, Elmi AH, Abderrahman AD**.** Inventory of plants used in traditional medicine in Somalia. IV. Plants of the families Passifloraceae- Zygophyllaceae. J Ethnopharmaco. 1993; 38(1): 18-19.

97. **Sangeeta D. et al,** Effect of *Tribulus Terrestris* on oxalate metabolism in rats, J Ethno pharmaco. 1994; 44(2): 61-6.

98. **Sarvin, B.**; Stekolshchikova, E.; Rodin, I.; Stavrianidi, A.; Shpigun, O. Optimization and comparison of different techniques for complete extraction of saponins from T. terrestris. J. Appl. Res. Med. Aromat. Plants 2018, 8, 75-82.

99. **Semerdjieva, I. B.**; Zheljazkov, V. D., Chemical Constituents, Biological Properties, and Uses of Tribulus terrestris: A Review. *Natural*

Product Communications **2019,** 14, (8), 1934578X19868394.

100. **Singh A,** et al. Involvement of the α1-adrenoceptor in sleep-waking and sleep loss-induced anxiety behavior in zebrafish. Neuroscience. 2013; 245:136147. [PubMed: 23618759]

101. **Song, X.;** Kong, J.; Song, J.; Pan, R.; Wang, L. Angelica sinensis Polysaccharide Alleviates Myocardial Fibrosis and Oxidative Stress in the Heart of Hypertensive Rats. Comput. Math. Methods Med. 2021, 2021, 6710006. 57.

102. **Stefanescu, R**.; Tero-Vescan, A.; Negroiu, A.; Aurica, E.; Vari, C.-E., A Comprehensive Review of the Phytochemical, Pharmacological, and Toxicological Properties of Tribulus terrestris L. *Biomolecules* **2020,** 10, (5), 752.

103. **Steenbergen PJ,** et al. The use of the zebrafish model in stress research. Prog. Neuropsychopharmacol. Biol. Psychiatry. 2011; 35:1432-1451. [PubMed: 20971150]

104. **Stewart, A.** M.; Braubach, O.; Spitsbergen, J.; Gerlai, R.; Kalueff, A. V., Zebrafish models for translational neuroscience research: from tank to bedside. *Trends Neurosci* **2014,** 37, (5), 264-78.

105. **Stewart AM,** et al. Perspectives of zebrafish models of epilepsy: what, how and where next? Brain Res. Bull. 2012; 87:135-143. [PubMed: 22155548]

106. **Stewart A,** et al. Modeling anxiety using adult zebrafish: a conceptual review. Neuropharmacology. 2012; 62:135-143. [PubMed: 21843537]

107. **Stewart, AM.;** Kalueff, AV. Developing better and more valid animal models of brain disorders. Behav. Brain Res. 2013. http://dx.doi.Org/10.1016/j.bbr.2013.12.024

108. **Stewart AM**, et al. Developing zebrafish models of autism spectrum disorder (ASD). Prog. Neuropsychopharmacol. Biol. Psychiatry. 2014; 50C:27- 36. [PubMed: 24315837]

109. **Sudheendran, A.**; Shajahan, M. A.; Premlal, S., A comparative diuretic evaluation of fruit and root of Gokshura (Tribulus terrestris Linn.) in albino rats. *Ayu* **2021,** 42, (1), 52-56.

110. **Sultana, R.;** Perluigi, M.; Butterfield, D.A. Lipid peroxidation triggers neurodegeneration: A redox proteomics view into the Alzheimer disease brain. Free Radic. Biol. Med. 2013, 62, 157-169. 54.

111. **Sun J.**, Wang H., Liu B. et al, "Rutin attenuates H2O2- induced oxidation damage and apoptosis in Leydig cells by activating PI3K/Akt signal pathways," Biomedicine & Pharmacotherapy, vol. 88, pp. 500-506, 2017.

112. **Vaváková M**. Duracková Z. and Trebatická J. "Markers of oxidative stress and neuroprogression in depression disorder," Oxidative Medicine and Cellular Longevity, vol. 2015, Article ID 898393, 12 pages, 2015.

113. Wang, S.; Kong, X.; Chen, Z.; Wang, G.; Zhang, J.; Wang, J., Role of Natural Compounds and Target Enzymes in the Treatment of Alzheimer's Disease. *Molecules* **2022,** 27, (13), 4175.

114. **World Health Organization**. Dementia [Fact Sheet]; 2023.

115. Yan, Z.-Q.; Chen, J.; Xing, G.-X.; Huang, J.-G.; Hou, X.-H.; Zhang, Y., Salidroside prevents cognitive impairment induced by chronic cerebral hypoperfusion in rats. *Journal OfInternational Medical Research* **2015,** 43, (3), 402-411.

116. **Yamazaki, H.**; Tanji, K.; Wakabayashi, K.; Matsuura, S.; Itoh, K. Role of the Keap1/Nrf2 pathway in neurodegenerative diseases. Pathol. Int. 2015, 65, 210-219.

117. **Yang, K.;** Chen, Z.; Gao, J.; Shi, W.; Li, L.; Jiang, S.; Hu, H.; Liu, Z.; Xu, D.; Wu, L. The Key Roles of GSK-3beta in Regulating Mitochondrial Activity. Cell Physiol. Biochem. 2017, 44, 1445-1459. 49.

118. **Yang, G**.-X.; Huang, Y.; Zheng, L.-L.; Zhang, L.; Su, L.; Wu, Y.-H.; Li, J.; Zhou, L.-C.; Huang, J.; Tang, Y.; Wang, R.; Ma, L., Design, synthesis and evaluation of diosgenin carbamate derivatives as multitarget anti-Alzheimer's disease agents. *European Journal of Medicinal Chemistry* **2020,** 187, 111913.

119. **Yin, Y.;** Li, H.; Chen, Y.; Zhu, R.; Li, L.; Zhang, X.; Zhou, J.; Wang, Z.; Li, X. Combination of Astragalus membranaceous and Angelica sinensis Ameliorates Vascular Endothelial Cell Dysfunction by Inhibiting Oxidative Stress. Evid. Based Complement. Altern. Med. 2020, 2020, 6031782. 56.

120. **Zhang Z. J**., "Therapeutic effects of herbal extracts and constituents in animal models of psychiatric disorders," Life Sciences, vol. 75, no. 14, pp. 1659-1699, 2004.

121. **Zhao, J**.; Tian, X.-C.; Zhang, J.-Q.; Li, T.-T.; Qiao, S.; Jiang, S.-L., *Tribulus terrestris L.* induces cell apoptosis of breast cancer by regulating sphingolipid metabolism signaling pathways. *Phytomedicine* **2023**, 155014.

122. **Zhu,** W.; Du, Y.; Meng, H.; Dong, Y.; Li, L., A review of traditional pharmacological uses, phytochemistry, and pharmacological activities of *Tribulus terrestris. Chemistry Central Journal* ***2017,*** *11, (1), 60.*

123. **Ziv L,** et al. An affective disorder in zebrafish with mutation of the glucocorticoid receptor. Mol. Psychiatry. 2013; 18:681-691. [PubMed: 22641177]

MIX
Papier aus verantwortungsvollen Quellen
Paper from responsible sources
FSC® C105338

Printed by Books on Demand GmbH, Norderstedt / Germany